INDIAN LOGIC AND ETHICS

INDIAN LOGIC AND ETHICS

Editor-in-Chief

Swami Chidatman Jee Maharaj

ANMOL PUBLICATIONS PVT. LTD.

NEW DELHI - 110 002 (INDIA)

ANMOL PUBLICATIONS PVT. LTD.
H.O.: 4374/4B, Ansari Road, Daryaganj,
New Delhi-110 002 (India)
Ph.: 23278000, 23261597
B.O.: No. 1015, Ist Main Road, BSK IIIrd Stage
IIIrd Phase, IIIrd Block,
Bangalore - 560 085 (India)
Visit us at: www.anmolpublications.com

Indian Logic and Ethics

First Published, 2009
ISBN 978-81-261-3627-8

PRINTED IN INDIA

Printed at Mehra Offset Press, Delhi.

Contents

Preface

India colourful and vibrant, a land as diverse as its people. A mosaic of faiths, cultures, customs and languages that blend harmoniously to form a composite whole. One of the world's oldest living civilizations—which gave to the world—the concept of zero, the primordial sound Aum..., Yoga and Buddhism.

She is old. She is young. The two diametrically opposite, different yet unique and dynamic faces of India coexist even at this turn of the millennium. With a history of heritage, learning and civilisation dating back to almost 5000 years or perhaps even more, the modern India presents itself as an ideal country. From snow clad mountains to breezy sparkling blue sea side. From arid deserts with shifting sand dunes to lush green country sides. From ancient relics and architectural ruins dating back into antiquity to modern cities, India has it all.

To the north it is bordered by the world's highest mountain chain, where foothill valleys cover the northernmost of the country's maximum states. Further south, plateaus, tropical rain forests and sandy deserts are bordered by palm fringed beaches. Its southernmost tip is the meeting point of three Oceans.

India is a land of staggering contrast with a mingling of the tradition and modernity that is an unique experience to be in India. One will forever remember the stay in this wonderful Homeland-a land truly of mystery and enchantment. Only India can offer such an astonishing variety of contrasts.

She has given birth to numerous spiritual leaders and founders of some of the popular religions of the present day world. India probably has the most religious diversity in any country. It's the birthplace of Hinduism, Buddhism, Sikhism and Jainism and is among the few places to have a resident

Zoroastrian population. India with her veritable treasure trove of culture, mysticism, philosophy, art, music and architecture to name a few. Much of India's classical music is devotional and the North Indian Hindustani and South Indian Carnatic streams are distinct and both have a complex 'raga' framework. The legacy of dance in India is tremendous. The classical dances of India are numerous such as Kathakali of Kerala, Bharatnatyam of Tamil Nadu, Kuchipudi of Andhra Pradesh, Manipuri and Odissi from Orissa are the prominent dance forms in this country that sways to an altogether novel beat. The Yakshagana, nautanki and puppetry are ancient folk forms that live on till date. The earliest specimens of Indian painting are the ones on the walls of the Ajanta Caves dating back to 2nd century BC. The Mughals had a huge impact on Indian art. The influence of Persian art brought placid garden scenes, illustrations from myths, legends and history into Indian art. Word craft, handicrafts, architecture and sculpture all contribute to this rich and varied domain. Indian literature, both in English and in the vernacular, is ever more popular around the world.

The book aims at presenting perspectives of different aspects of Indian culture. It has topics on Indian classical music and dance, painting, ayurveda and yoga which are traditional and indigenous system of physical and mental wellbeing, language, philosophy and excellence achievements by Indian in the different branches of science and social sciences. It also includes our ancient religious and classic literature.

It is hoped that this book will be welcomed by the general readers as also scholars who are interested in the Indian heritage.

—*Editors*

1

Hindu Ethics

Souls and Matter are only the body of God. Creation is a real act of God. It is the expansion of intelligence. Matter is fundamentally real and undergoes real revelation. The Soul is a higher mode than Matter, because it is conscious. It is also eternally real and eternally distinct. Final release, that comes, by the Lord's Grace, after the death of the body is a Communion with God.

This philosophy believes in liberation through one's Karmas (actions) in accordance with the Vedas, the Varna (caste or class) system and the four Ashramas (stages of life), along with intense devotion to Vishnu. Individual Souls retain their separate identities even after moksha. They live in Fellowship with God either serving Him or meditating on Him. The philosophy of this school is Sri Vaishnavism, a branch of Vaishnavism.

Interpretation of Mahavakyas

All Vedantic schools need to substantiate the meaning espoused by Mahavakyas which occur throughout Upanishadic literature. The interpretation of these *Grand Pronouncements* serve as the cornerstone for establishing each school of thought. The most significant among them is:

Translated literally, this means *All this is Brahman*. The ontology of Vishishtadvaita system consists of:

1. Ishvara is *Para-brahman* with infinite superlative qualities, whose substantive nature imparts the existence to the modes.

2. Jivas are *chit-brahman* or sentient beings (which possess consciousness). They are the modes of Brahman which show consciousness.
3. Jagat is *achit-brahman* or matter/Universe (which are non-conscious). They are the mode of Brahman which are not conscious.

Brahman is the composite whole of the triad consisting of Ishvara along with his modes *i.e.* Jivas and Jagat. Hence, "all this is Brahman" denotes the triad of entities.

1. Sarvam khalv idam brahma from Chandogya Upanishad 3.14.1
2. Ayam atma brahma from Mandukya Upanishad 1.2

Translated literally, this means the *Self is Brahman*. From the earlier statement, it follows that on account of everything being Brahman, the self is not different from Brahman.

3. Tat tvam asi from Chandogya Upanishad 6.8.7

Translated literally, it means *Thou art that*

that here refers to *Brahman* and *thou* refers to *jiva*

The vakya establishes the identity of the jiva and Brahman. The issue here is if the identity involves establishing a unique identity or a universal identity. The difference is as follows:

1. *Unique Identity:* Atman is Brahman; Nothing else is Brahman; Brahman is reality and therefore everything else is illusion
2. *Universal Identity:* Atman is Brahman in the same way as everything else is Brahman.

Ramanuja chooses to take the position of universal identity. He interprets this passage to mean the subsistence of all attributes in a common underlying substratum. This is referred to as *samanadhikaranya*.

Thus Ramanuja says the purport of the passage is to show the unity of all beings in a common base. Ishvara (*Parabrahman*) who is the Cosmic Spirit for the pan-organistic body consisting of the Universe and sentient beings, is also simultaneously the innermost self (Atman) for each individual sentient being (Jiva).

All the bodies, the Cosmic and the individual, are held in an adjectival relationship (aprthak-siddhi) in the one Isvara.

Tat Tvam Asi declares that oneness of Isvara.

When multiple entities point to a single object, the relationship is established as one of substance and its attributes.

For *e.g.* in a statement:

Jack is a tall and intelligent boy

The descriptors *tallness*, *intelligence* and *boy-ness* all refer to a common underlying *Jack*.

Similarly, when the upanishads declare Brahman is the Universe, Purusha, Self, Prana, Vayu, and so on, the entities are attributes or modes of Brahman.

If the statement tat tvam asi is taken to mean as only *the self is brahman*, then *sarvam khalv idam brahma* will not make sense.

Understanding Neti-Neti

This is an upanishadic concept which is employed while attempting to know Brahman. The purport of this exercise is understood in many different ways and also influences the understanding of Brahman. In the overall sense, this phrase is accepted to refer to the indescribable nature of Brahman who is beyond all rationalisations. All descriptions of such an entity will necessarily have to be partial or fall short of the actual.

The typical interpretation of Neti-Neti is *not this, not this* or *neither this, nor that*. In Vishisht Advaita, the phrase is taken in the sense of *not just this, not just this* or *not just this, not just that*. This means that Brahman cannot be restricted to one specific or a few specific descriptions. Consequently, Brahman is understood to possess infinite qualities and each of these qualities are infinite in extent.

Purpose of Human Existence

The purpose or goal of human existence is called as Purush Artha. According to the Vedas, there are four goals namely Artha (wealth), Kama (pleasure), Dharma (righteousness) and

Moksha (permanent freedom from worldly bondage). According to this philosophy, the first three goals are not an end by themselves but need to be pursued with the ideal of attaining Moksha.

Moksha

Moksha is a state where the jiva achieves oneness with Brahman in terms of all knowership and possessing qualities free from all worldly evils and defects. The jiva however doesn't possess the power to manifest/create and unmanifest/destroy. Neither does it have the power to grant Moksha.

The union of Atman and Brahman is likened to a situation where tiny lamps come under the blaze of the Sun. The lamp and Sun are still identifiable as different sources of light and yet the light arising from them is indistinguishable.

Moksha doesn't involve destruction of the self ("I") consciousness of the jiva.

Comparison with Western Non-dualism

Baruch Spinoza, the 17th century rationalist philosopher, in his magnum work Ethics establishes the nature of GOD. Spinoza's pan-organistic God (*i.e.* God revealed as orderly nature) is comparable to Brahman (having the individual selves' and Universe as its body)

Spinoza makes the following propositions on the nature of God, in his work "Ethics". These positions closely reflect the Vishist Advaitic position on the nature of Brahman:

Proposition xi. God, or substance consisting of infinite attributes, of which each expresses eternal and infinite essentiality, necessarily exists.

Proposition xv. Whatsoever is, is in God, and without God nothing can be, or be conceived.

Proposition xvii. God acts solely by the laws of his own nature and is not constrained by anyone.

Proposition xviii. God is the indwelling and not the transient cause of all things.

Proposition xix. God and all the attributes of God are eternal.

Proposition xxx. Intellect, in function finite, or in function infinite, must comprehend the attributes of God and the modifications of God, and nothing else.

Vishisht Advaita and Sri Vaishnavism

The Absolute Supreme Reality referred to as Brahman, is a Transcendent Personality with infinite superlative qualities. He is Narayana, also known as Lord Vishnu. He is also the other two members of the Trimurti, namely, Creator Brahma and Shiva, the Lord of Deluge.

A man who has discrimination for his charioteer and holds the reins of the mind firmly, reaches the end of the road; and that is the supreme position of Vishnu.-1.3.9 Katha Upanishad.

Beyond the senses are the objects; beyond the objects is the mind; beyond the mind, the intellect; beyond the intellect, the Great Atman; beyond the Great Atman, the Unmanifest; beyond the Unmanifest, the Purusha. Beyond the Purusha there is nothing: this is the end, the Supreme Goal.-1.3.10,11 Katha Upanishad.

In terms of theology, Ramanujacharya puts forth the view that both the Supreme Goddess Lakshmi and Supreme God Narayana together constitute Brahman-the Absolute. Sri Lakshmi is the female personification of Brahman and Narayana is the male personification of Brahman, but they are both inseparable, co-eternal, co-absolute and are always substantially one. Thus, in reference to these dual aspects of Brahman, the Supreme is referred to in the Sri Vaishnava Sampradaya as Sriman Narayana.

Conclusion

Vishishtadvaita conclusions may be briefly summarised as below:

Narayana is the Absolute God. The Soul and the Universe are only parts of this Absolute and hence, Vishishtadvaita is panentheistic. The relationship of God to the Soul and the Universe is like the relationship of the Soul of Man to the body of Man. Individual souls are only parts of Brahman. God, Soul and Universe together form an inseparable unity which is one and has no second. This is the non-duality part. Matter and Souls inhere in that Ultimate Reality as attributes to a substance. This is the qualification part of the non-duality.

Vishishtadvaita philosophy provided the philosophical basis for the establishment of Sri Vaishnavism and gave Vedantic backing to the brimming devotion of the Alwar saints and their composition of wonderful poetry and devotional songs in praise of Lord Vishnu. The succession of great Master-Expositors and spiritual giants of Vishishtadvaita school starts with the twelve Alwars, who left behind an imperishable legacy of Tamil devotional poetry in the form of 4000 songs, now called the Nalayira divya prabandham.

The Hinduism—Legends and Customs—Lord Brahma

We all think we know the world we are living in Jet, we all know in through our own culture's eyes, and it is a known thing that culture is strongly influenced by religion. And one of the oldest religions of this world is the Hindu religion.

In Hinduism, everything begins, ends and it is included in ohm,, which is a sign representing the trinity of God. Like the Christians, the Hindu God is divided in 3 different gods, which are one and the same person at the same time. It's the same way a person can be somebody's brother, somebody's father and somebody else's husband, but being at the same time one single person. The gods Brahma, Vishnu and Shiva represent the Hindu trinity.

Lord Brahma is the Generator, the creator of earth and everything there is: animals, plants and humans. Even being the generator of all things, he isn't worshipped because he married his sister.

Lord Vishnu, the most worshipped one, is the Operator. He is one that makes all the rules of the society, the one that makes things move. He usually uses material thing to persuade people to do the different things he wants them to do. Lord Vishnu is the most handsome one and he is usually represented like a beautiful young man with 4 arms or like a normal beautiful young man that has the *shakra* in his right hand and a snail shell from which it flows sand in the left one. The snail shell and the sand represent the flow of the time.

Lord Shiva is the Destroyer, the most important of the 3 gods. He is the one who will arise the *Divyajyoti*, the Hindu "apocalypse", the end of the world. Hindus believe in reincarnation, so the ones that will go to haven and the ones who will go to hell are the ones who made good deeds and bad deeds during their life until the *Divyajyoti*'s time when there will be too much sin on this world. When this will happen, he will open his third eye, and the world will burn. He is represented usually in the meditation position, with his third eye opened that he gained through meditation. That's why Hindu's wear a coloured dot between their eyes. It represents worshipping the intellect, the power of reason and will. Lord Shiva has an eye instead of a dot because he took the meditation to such level that a third eye opened, the eye of the mind.

Actually, it is with his third eye that he killed *Kaam Dev*, the god of love, because it was affecting him too much.

There are many gods in the Hindu religion. Everything that exists has it own got. There are about 63.000 gods. Some of the most known are *Varun Dev*, the god of air, *Arun Dev*, the sun god, *Agni Dev*, the god of fire and *Kaam Dev*, the god of love, who was killed by lord Shiva. It is useless to pray to him because he's dead. Still, there is a mantra that can invoke *Kaam Dev* into one's life and ask for his help. Mantras are prayers addressed to different gods.

There are many legends in the Hindu mythology and many of them have similarities with the ancient Greek myths. One of them is when devils attacked the haven, the way the titans attacked Olympus. When this happened, *Indra*, the king of haven, went to ask Shiva what to do because the world was in danger. And Shiva told him that only by throwing into the devils with the bones of a holly man the devils would die. And this holly man vas *Dadhichi*, the yoghin. *Indra* left to find *Dadhichi* and the found him meditating. His level of meditation was so high that he was able to make his soul leave his body. So, the gods could take his bones and throw them into the devils, thus saving the world, thanks to the old man's sacrifice.

It is a known thing that the woman is seen as the most fragile and delicate thing is all the societies of the world. Still, within the woman lies a great power. A power big enough to save the world, as Hindus say. When the world is in real danger, the 3 divine powers: the Generator, the Operator and the Destroyer, *Brahma*, *Vishnu* and *Shiva*, join together, thus generating 4 powers, That take the shape of 4 women, 4 goddesses: *Durga*, *Kali*, *Saraswati* and *Lakshmi*. Goddess *Durga* is the enemy of all the evil and the bad things in the world. The ones who worship *Durga* never forgive their enemies. *Kali* is the some like *Durga*.

She is the enemy of all the evil in the world and she is a destroyer, like *Shiva*, but she only destroys the evil. She's represented as being completely black, like the earth, whose rage she represents, wearing a garland of human skulls and she has her tongue always outside her month. She can also be invoked in black magic practices, but she is one of the 4 powers that can save the world in time of need. When the mother earth is angry, she takes the shape of goddess *Kali* to revenge herself. The image of *Kali* presented in the movie "Indiana Jones and the Temple of Doom" is completely fake and it has nothing to do with goddess *Kali*. Actually, the Indian people took the movie as in insult and it was forbidden in India. *Saraswati* is the goddess of intelligence and knowledge. Anybody who is student or learning something should worship *Saraswati* and ask her to make them keep their intelligence in their heads, unless it will vanish. They should ask her for the power of the concentration and of learning.

And finally, *Lakshmi*, who is the goddess of wealth and money. These are the 4 women, the 4 powers generated by the union of the 3 divine lords: *Brahma, Vishnu* and *Shiva*.

Unlike other religions, Hindus are born into Hindu religion. In Christianity, we baptise our children in the church of Jesus. Muslims and Jews circumcised their children in order to make them members of the Muslim or Jewish religious society. Unlike these, Hindus are born into their religion. They are members of the religious society strait from birth. They don't need a consecration like we need the baptise. All the children, regardless of sex, receive their mother's name until she decides to give them a name around the age of 4 years when the legal papers are completed as well.

The marriage ritual is also very different. There are 3 types of ceremony: in the jungle, in the society and in the temple. The ceremony that takes place in the jungle, is secret, only the bride and the groom know what happens. It is forbidden now and considered illegal because it is thought to be a devil's ceremony, due to its secret nature. The most common one is the social marriage but the rituals are very different depending on religions. India is a very big country and big countries mean many different traditions. It is the girl's parents that go to ask the boy's hand in marriage from his parents. The ceremony has 3 stages, which happened during in a night. In the first phase, the girl is taken from her house where she waits for the groom to come for her. She stays on a throne, wearing a red dress.

During the night, the bride, the groom and the bride's parents can't eat anything. When they arrive at his place, she changes her dress and wears a yellow one. A wood fire is set in the house. The girl's parents put her hand into the groom's hand, they are tied, and they surround the fire 7 times. Each time they surround the fire they take an oath. The first 4 times, the man is leading the woman and he takes the first 4 oaths. Than the last 3 times, the woman leads the man and it is her who takes the last 3 oaths. He swears always to protect her, to take all responsibility of her, never to let her cry, and she swears to let death take her first and also to make him happy, and take care of him. Than there is a feast with food and music and dance where everybody participates and she wears a red dress again.

Finally the ritual of death. If the one who dies is a child, a single girl or a single boy, or somebody who died because of a snakebite, he or she is drowned into a sacred river. All the rivers are believed to be sacred because all of them have their own god. If the one who died was married he or she is dressed as a groom or a bride and they are burned on a pyre. When the hair has burned and they can see his skull, they break it with a stick named *copalcria* to release the soul of the dead from its body. Women are not allowed to witness the ceremony. If the dead is a man, the one who will release the soul is his father, his eldest son or his brother. If the dead is a woman, it will be her husband or her eldest son. After the body is burned, there is 13 days ceremony. After 10 days, the men of the family shaves their heads in order to purify the house which isn't pure any more because of the presence of the dead's soul. They say that the soul of the dead stays around the house until all the 13 days ceremonies performed. In the 13th day, the family organises a feast for the *Brahmas*, and after this, the ceremony is completed and the soul is free to go.

The *Brahmas* are the upper class of the Indian society classes that are in number of 4: *brahma*, *shatria, veshia* and *shutra.* In the ancient times, people could change their cast if they had the necessary skills, but nowadays this isn't possible any more: the cast you are born it is like your identity card and you can not change it. The *brahmas* are the scholars, the ones that have the knowledge, the ones that have to teach the people. *Kshatriyas* are the warriors, the ones that fight to protect the country. *Veshias* are the traders, while *shutras* are the lower cast, the ones that serve the upper ones.

Hindus also have their holly days. Actually, each day is a different celebration. But there are four very important: *Holy, Rakshbandhan, Dusshera* and *Diwali.*

Holy is celebrated in March, depending on the position of the moon. It is the festival of colours, and it signifies the victory of love over hatred and the victory of good over evil. They say that once there lived a devil named *Hiraneakashiap* and in his kingdom nobody was allowed to worship God. But his son, *Prahalad*, was a great worshipper of God. His father was so upset with his son, that he tried to kill him. *Prahalad* had an aunt, his father's sister, whose name was *Hulika* and she was also a devil. She had a magic shawl that protected against fire. They tried to set a trap for *Prahalad* and *Hulika* stood in the middle of the fire, with her shawl on. But, by God's will, the shawl passed into *Prahalad*'s hand and the one who died in the fire was *Hulika*. *Hiranyakashyap* tried to kill his son again and again, until one day he put him inside a column and asked him: "where is your God now?".

The devil was protected by a magic that he couldn't be killed by an animal, a human or a god, not on earth and not on air. So, when he asked:" where is your God? I can see no God here?", God came as a creature, half human, half lion, a sphinx. He put the devil on his knees, leaving his feet on the ground. He ripped him apart with his claws from his waste. So, no animal, no god and no human killed him, on the earth and not in the air, 'cause he was both on the earth and in the air, and at the same time not on the earth and not in the air. The name of the festival, *Holy*, comes from *Prahalad's* aunt, *Hulika*.

Rakshabandhan is the festival of brothers and sisters. It is in august and it is the most important Hindu festival. The sisters tie a silk band on the wrest of their brothers, and the brothers swear to protect their sisters of everything that's bad in the world. The wife of *Indra*, the king of heaven first performed the ritual, when the heaven was attacked by devils.

Dusshera takes place in September, October and November, depending of the position of the moon. It is the victory of good over evil. Its meaning is the victory of prince *Rama,* the incarnation of *Lord Vishnu* over the devil *Ravaṇa.* He killed the devil and freed his wife and his kingdom, Lanka.

The last one of the four festivals is *Diwali* festival, the festival of light. It takes place nineteen days after the *Dusshera* festival. After prince *Rama* conquered Lanka and freed *Ravana's* wife, when he arrived back to whose own kingdom, after nineteen days, the people of his city enlightened the town with oil lamps in his honour. The main idea of Hinduism is that we have to do what we think is right, to work to live a good life, the way the holy books teach us and don't expect anything and not think about the consequences of our deeds. We should leave the rest to God, accept his wish and accept our destiny.

2

The Hindu Theory of World Cycles

In the Light of Modern Science

Traditional Hindu scriptures view history as cyclical in character, with vast repeating series of ages. Each age has its own particular qualities.

Interestingly, this system seems to be taken literally by modern Yoga masters such as Swami Muktananda, Baba Hari Dass, Swami Vishnu Tirtha, and so on. As a Western devotee, I found myself wondering exactly what the full system of Hindu cycles is, and how it relates to the findings of modern science. Following is my attempt to explain what I found out.

Traditional Puranic Model

The Hindu Puranas describe a number of cycles within cycles. Discussions of these cycles can become confusing because different cycles are measured in different types of units. For example, the cycles are often described in units of deva years, each of which equals 360 human years.

The following description starts with the smaller cycles and works up to the larger ones. The length of each cycle is given in ordinary human (earth) years, as well other units where appropriate.

Large numbers are described using the conventions of American English: thus, a million is a thousand, a billion is a thousand million, a trillion is a thousand billion.This description is based on numerous sources, which are given in the "References" section at the end of this document.

Maha Yugas

The smallest cycle is called a maha yuga. A maha yuga is 4,320,000 human years. Each maha yuga is subdivided into the following four ages, whose lengths follow a ratio of 4:3:2:1:

Satya Yuga (also called Krita Yuga)

This first age is 1,728,000 human years. Also known as the Golden Age or age of Truth. The qualities of this age are: virtue reigns supreme; human stature is 21 cubits; life-span is a lakh of years, and death occurs only when willed.

Treta Yuga

This second age is 1,296,000 human years. Also known as the Silver Age. The qualities of this age are: the climate is three quarters virtue and one quarter sin; human stature is 14 cubits; life-span is 10,000 years.

Dvapara Yuga

This third age is 864,000 human years. Also known as the Bronze Age. The qualities of this age are: the climate is one half virtue and one half sin; life-span is 1,000 years.

Kali Yuga

The fourth and last age is 432,000 human years. Also known as the Iron Age. This is the age in which we are presently living. The qualities of this age are: the climate is one quarter virtue and three quarters sin; human stature is 3.5 cubits; life-span is 100 or 120 years.

Toward the end of a Kali Yuga, various calamities cause a good deal of destruction. Baba Hari Dass states that creation disappears at the end of a Maha Yuga and remains in seed form inside Brahma. However, other sources do not suggest anything so drastic; it is possible that Hari Dass was really thinking of the end of Brahma's daytime or Brahma's life when he wrote this description.

Brahma Days (Kalpas)

A kalpa is a single daytime period in the life of Brahma, the creator god. Two kalpas are a day and a night of Brahma.

Each kalpa is composed of 1,000 maha yugas. A kalpa is thus equal to 4.32 billion human years.

At the end of Brahma's daytime period, the Three Worlds (Bhuloka, Bhuvarloka, Swarloka) and the seven underworlds (of the nagas) are temporarily dissolved (pralaya); that is, the same folks can be reincarnated when the next day of Brahma begins.

The Vishnu Purana states that at the end of the daytime period of Brahma, a dreadful drought occurs that lasts 100 years, and all the waters are dried up. The sun changes into seven suns, and the three worlds (Bhurloka or Earth, Bhuvarloka or the lowest heaven, and Svargloka or the next higher heaven) and the underworlds are burned bare of life. The inhabitants of Bhuvarloka and Svargloka flee to the next higher heaven, Maharloka, to escape the heat; and then to the next higher heaven, Janaloka.

Then mighty clouds form and the three worlds are completely flooded with water. The lord Vishnu reposes on the waters in meditative rest for another whole kalpa (4.32 billion years) before renewing the creation.

The destruction that takes place at the end of a daytime of Brahma is referred to as naimittika, which is incidental or occasional. The characteristic of this destruction is that the three worlds continue to exist but are made uninhabitable. The souls of individuals also continue to exist to be reincarnated in the next daytime of Brahma.

Brahma Years

A year of Brahma is composed of 360 day/night cycles of Brahma, or 720 kalpas, or 8.64 billion human years.

Brahma Life

The life-span of Brahma is 100 Brahma years, or 72,000 kalpas, or 311.04 trillion human years.

At the end of the life of Brahma, all worlds are completely dissolved (mahapralaya). No one is reincarnated from these worlds ever again.

Manvantaras

Another cycle that overlaps the others is that of manvantaras. Each kalpa is reigned over by a succession of 14 Manus, and the reign of each Manu is called a manvantara. A single manvantara is approximately 71 maha yugas.

Coomaraswamy states: "Each Manvantara is followed by a Deluge, which destroys the existings continents and swallows up all living beings, except the few who are preserved for the repeopling of the earth."

Our Position in History

We are located in the fifty-first Brahma year of the life of our Brahma.

Within that Brahma year, we are in the first Brahma day, called the Varaha kalpa.

Within that Brahma day, we are in the seventh manvantara, and in the 28th maha yuga of that manvantara. This would place us at about the 454th maha yuga of the 1,000 maha yugas that comprise this day of Brahma.

Within this maha yuga, we are in Kali Yuga. The 5100th year of Kali Yuga will correspond to the year 2,000 AD. That means that we are fairly early in Kali Yuga and this age will continue more than 426,000 more years.

Variant Interpretations of Hindu Chronology

The "Traditional Puranic Model" described above is agreed upon by most authors on Hinduism and Yoga. Six different authors, listed at the end of this paper, describe this model identically.

However, several other authors, some of them well-known Hindu teachers, have published descriptions of the cycle of ages that differ from the traditional Puranic model. These variant theories are described below.

Sri Yukteswar

In the introduction to his book *The Holy Science*, Sri Yukteswar describes an interesting variant of the Hindu theory of ages. According to him,...the sun, with its planets and their moons, takes some star for its dual and revolves around it in about 24,000 years of our earth-a celestial phenomenon which causes the backward movement of the equinoctial points around the zodiac. The sun also has another motion by which it revolves round a grand centre called *Vishnunabhi*, which is the seat of the creative power, *Brahma*, the universal magnetism. *Brahma* regulates *dharma*, the mental virtue of the internal world.

Yukteswar goes on to explain that the sun's 24,000 year revolution around its companion star takes the sun progressively closer, and then progressively further away from the mystic centre Vishnunabhi. In his system, dharma increases as we approach Vishnunabhi and decreases as we draw away from it. The cycle of yugas takes place twice in each 24,000 year revolution. As the sun recedes from Vishnunabhi, the ages pass in the usual order: Satya, Treta, Dvapara, Kali. As the sun approaches Vishnunabhi, the ages pass in the opposite order: Kali, Dvapara, Treta, Satya.

The length of the yugas is: Satya Yuga, 4800 years; Treta Yuga, 3600 years; Dwapara Yuga, 2400 years; and Kali Yuga, 1200 years. The yugas during the approach to Vishnunabhi he calls 'ascending' yugas; those during the retreat from Vishnunabhi he calls 'descending' yugas. The most recent ascending Kali Yuga began in 499 AD. Since 1599 AD we have been in the ascending Dwapara Yuga, with consequent advances in human culture and knowledge.

Yukteswar goes on to say that the Hindu almanacs, which correspond to the traditional Puranic model described previously, are in error. The error crept in during the dark years of Kali Yuga when scholars misinterpreted the scriptures. Regarding the conventional view that we are currently in Kali Yuga, Yukteswar says flatly that it is not true.

Yukteswar's model thus differs from the traditional in the following respects:

- A cycle of four yugas takes 24,000 years instead of 4,320,000.

- The yugas alternate between ascending and descending trends instead of always proceeding in the same order. This alternation becomes necessary once you posit that the ages result from our changing distance from Vishnunabhi, rather than a deliberate divine intervention at the end of Kali Yuga.
- The greater cycles like kalpas, manvantars, and life-span of Brahma go unmentioned.

Paramahansa Yogananda

Paramahansa Yogananda was a disciple of Sri Yukteswar and one of the best-known Hindu teachers ever to visit the West. He wrote the perrenial bestseller *Autobiography of a Yogi*.

In the latter book, Yogananda describes and endorses Yukteswar's theory of world cycles. However, in a footnote, Yogananda adds the following:

The Hindu scriptures place the present world-age as occurring within the Kali Yuga of a much longer universal cycle than the simple 24,000 year equinoctial cycle with which Sri Yukteswar was concerned. The universal cycle of the scriptures of 4,300,560,000 years in extent, and measures out a Day of Creation. This vast figure is based on the relationship between the length of the solar year and a multiple of pi (3.1416, the ratio of the circumference to the diameter of a circle).

The life span for a whole universe, according to the ancient seers, is 314,159,000,000,000 solar years, or "One Age of Brahma." The Hindu scriptures declare that an earth such as ours is dissolved for one of two reasons: the inhabitants as a whole become either completely good or completely evil. The world mind thus generates a power that releases the captive atoms held together as an earth.

This statement seems at first to reconcile Yukteswar's theory with the traditional view, but in fact actually contradicts both.

Regarding Yukteswar's theory, in his own writing he clearly states that the traditional Hindu almanacs are in error and suggests how the error came about. He states that the length of the yugas, given in ordindary years in the scriptures, were misinterpreted by later scholars as being counted in units of "deva years" which are much longer. This method led the scholars to believe that the yugas are much longer than they really are. Yukteswar's theory is thus clearly intended to replace, not to supplement, the traditional interpretation.

Regarding the "much longer universal cycle" that Yogananda describes, he states that a Day of Creation is 4,300,560,000 years. This is close but not identical to the traditional number, which is 4,320,000,000 years. Similarly, Yogananda gives 314,159,000,000,000 years the the life of Brahma, whereas traditionally the number is slightly different: 311,040,000,000,000 years.

It seems likely that Yogananda arrived at these different figures because he wanted to derive them from some multiple of the ratio pi. The exact manner in which pi enters into the calculation is, unfortunately, not explained in his footnote.

Yugas and Science

If we restrict our attention to the traditional interpretation, we find that it makes a number of significant predictions that can be compared with the findings of modern science.

Great Culture Preceded Us

Beginning about 3,894,000 years ago, there is supposed to have been a great civilisation in which people were happier, taller, and much longer lived than they are today.

By contrast, scientists currently believe that homo sapiens evolved from more primitive forbears about 300,000 to 400,000 years ago. Humanity is supposed to have domesticated plants about 12,000 years ago, and animals shortly thereafter. Prior to that point, humans are thought to have been hunter-gatherers and possibly scavengers.

The scientific view is based on fossil evidence. The mystery is how an advanced civilization posited by the Hindu theory could have vanished without leaving any trace for us to find.

Cyclic Catastrophes

A number of periodic catastrophes are described in the Hindu scriptures:

- At the end of each Kali Yuga, some type of destruction takes place. The most recent instance would be a bit under 4 million years ago. Other instances should be found every 4,320,000 years before that.
- At the end of each manvantara, a great flood wipes out most life on earth. The most recent occurrence would be 120,534,000 years ago. Other instances would occur at intervals of 306,720,000 years.
- At the end of the kalpa, all life on earth is destroyed. The most recent instance would be 2,267,574,000 years ago.
- At the end of a life of Brahma, the entire universe and all its heavens and hells are destroyed. The current universe would have come into existence more than 150 billion years ago.

Interestingly, current scientific research does support the existence of occasional mass extinctions in earth's history. These include the following major extinctions:

- At the Ordovician/Silurian transition, about 425 million years ago.
- Near the Devonian/Carboniferous transition, about 345 million years ago.
- At the Permian/Triassic transition, about 230 million years ago.
- At the Triassic/Jurassic transition, about 180 million years ago.
- At the Cretaceous/Tertiary transition, about 63 million years ago.

Additionally, some scientists have identified what they believe is a cycle of periodic mass extinctions occurring every 26 million years.

Unfortunately, none of these specified dates corresponds to the catastrophes called for by the Hindu theory.

Current scientific estimates of the age of the universe range from 7 billion to 20 billion years. By contrast, the Hindu theory calls for a universe more than 150 billion years old.

On the other hand, the dates ascribed by scientists to the various geologic periods have been revised a number of times on the basis of new evidence, and could possibly be revised again in the future. Further, cosmology could reasonably be described as an infant science, and the age it ascribes to the universe as a whole might also change on the basis of new theories and evidence.

Downward Trend

By far the greater divergence from modern science is in the overall pattern of the Hindu theory. The Hindu and scientific patterns differ in the following ways:

- The main pattern in the Hindu theory is one of cycles.
- In the Hindu theory, life follows a downward trend most of the time, from the finest age to the worst. At the end of the worst age, Kali Yuga, divine intervention rapidly destroys the wicked and restores everything to its pristine state.
- In the Hindu theory, humanity is always present. The concept of evolution is confined to spiritual evolution; that is, each soul takes life in a series of lower to higher animal forms before finally incarnating as a human being.

By contrast, the fossil record of life on earth indicates that life began with very simple forms and later developed more complex organisms. The advent of humanity appears to be an extremely recent development when compared to the history of life on earth. Humanity itself does not appear to have existed long enough to have participated in the vast cycles of ages posited by Hindu theory.

Conclusion

There is no scientific support for the Hindu theory of world cycles. Further, current scientific theory contradicts Hindu theory in many respects. It is best to begin by acknowledging this truth, as such an acknowledgement can form the basis for interesting discussions of the different *ways of knowing* that underly the more specific differences.

Esoteric and Exoteric Knowledge

The Hindu religion is a vast ocean of religious thought, springing as it does from the earliest times, long before the dawn of history, and comprises in its multi-coloured texture shade after shade, an endless variety of design and pattern as it grew in the human mind; from animism to Nature worship, from powers of Nature in the abstract to personified and concretized natural forms, from gods and goddesses to the one Supreme God, first personal, and then impersonal; from form to formless.

The Hindu Pantheon offers a view of a vast and mighty host to the curious inquirer who pierces into the mists of the hoary past. Heliolatry, the worship of Helios or the Sun, was a common practice with the people of the world. Sol or Sun has ever been an object of great veneration for man and has been adored and worshipped all the world over from time immemorial. The ancient Greeks and the Romans built temples to Apollo or Phoebus as they termed the Sun-god in their own time. In all their temples, the image or representation of the Sun-god occupied an important place in their hierarchy. There is a famous Sun-temple in Konark, in South India, and in the historic town of Mooltan or the land of the Sun, in the North. In addition, Jog-maya or Jot-maya temples dot the whole Indian subcontinent.

The ancient Greeks also spoke of Shabd. It is written of Socrates that he heard within him a peculiar sound which pulled him irresistibly into higher spiritual realms. Pythagoras also talked of Shabd, for he described God as "Supreme Music of the nature of Harmonies." God was to him, "Absolute Truth, clothed in light." When he commanded an eagle to fly down to him and a bear to stop ravaging Apulia, the wondering multitude inquired of him the source from where such powers came to him. He replied that he owed it all to the "Science of Light."

Again, in the Greek language we have the mystical word Logos. It comes from the root lego which means to speak and from it we have the common terms, monologue, dialogue, prologue, epilogue and so on. The Logos means and stands for the "Word" or "Reason." The term Logos also occurs in both Hebrew and Christian philosophy and theology and is used, in its mystic sense, by the Hellenistic and Neo-platonist philosophers.

The Christians use it to denote the second person of the Trinity. The ancients in the West inherited this concept from their ancestors who, thousands of years before the Christian era, had come to acquire a great love and adoration for Surya which they regarded as the be-all and the end-all of all human endeavours in their search for the mighty power of God, and as a visible representation to this earth. They carried this notion along with them wherever they went, Eastward or Westward, and composed hymns and chanted psalms in praise of the glorious orb, the source of all life on this solar planet. Those who settled in Iran (Persia) and came subsequently to be known as Parsees, still worship the great deity in yet another form-fire-which they keep burning all the time in their temples as symbolic of the sacred flame that burnt in the human heart and always sprang heavenward. Ratu Zoroaster, the Iranian prophet of life and light, sang in loving and living faith of the greatness of the God of Light and taught the people to do so.

Agni or fire was a hidden secret with the gods, who guarded this mysterious power very jealously. It was, as the Greek legend goes, stolen by Prometheus and given to man, for which Jupiter, the father god, bound him to eternal torture. In Chapter VI of the Chhandogya Upanishad, it is said to be "the prime element whose creation made possible that of other elements, water, earth," etc.

The second branch of the Aryans which turned eastward into the Indo-Gangetic plain also referred lovingly to Aditya; and we have hymns in the Vedas addressed to Hiranyagarbha, Savitar and Usha, all of which stand for the One life-sustaining power, the Sun. The worshipful Masters of the Vedic age were, one and all, admirers of the purifying and healing attributes of the Sun-god, and so no wonder that we see many hymns in the Vedic literature deifying the sun. In Book X, 121, we find:

In the beginning rose Hiranyagarbha, born as the only lord of all created beings;

He fixed and holdeth up this earth and heaven;

What god shall we adore with our oblation?

What time the mighty water came, containing the Universal germ, producing Agni,

Thence sprang the God's One Spirit into being:

What god shall ule adore with our oblation?

In another hymn, he is referred to as "the self-radiant wise Aditya.; In Book I, 113, we have a hymn to Dawn and in it occur, inter alia, the following lines:

This light is come, amid all lights the fairest; born is the brilliant, far: extending brightness.

Night, sent away for Savitar's uprising, hath yielded up a birthplace, for the morning...

Arise! the. breath, the life, again hath reached us: darkness hath passed away, and light approacheth.

She for the sun hath left a path to travel; we have arrived where men prolong existence.

All this could be taken on the literal plane as little more than Nature-worship, an adoration of the sun, understandable among a people dependent upon agriculture for their existence. But ancient Indian literature has an elusive quality. It seems to teach us at one level, and when we have adjusted ourselves to it, it suddenly shifts us to another. He who can follow its subtleties finds in it a richness rarely to be met elsewhere.

There is multiplicity of meanings, ranging from the physical to the cosmic and the spiritual, and from the literal to the symbolic and esoteric, which challenge us at multiple levels of experience and offer us worthwhile rewards. Thus, when we begin studying these frequent references to the sun, we begin to see that the "sun" referred to is not always the centre of our physical Universe, which we initially took it to be. Thus,in the Isha Upanishad, we are told:

The door of the True One is covered with a golden disk.
Open that, O Pushan, that we may see the nature of the True One.

After recounting such statements, when we read of Brahman or the Supreme One, as being Jyotisvat, full of light, and Prakashvat, endowed with splendor, we begin to discover in such terms an esoteric significance we earlier overlooked. This comes to a head when we read the Gayatri, the tenth mantra of the sixteenth sutra in the third mandala of the Rig Veda:

Muttering the sacred syllable," Aum" rise above the three regions,
And turn thy attention to the All-Absorbing Sun within.
Accepting its influence be thou absorbed in the Sun,
And it shall in its own likeness make thee All Luminous.

This mantra is considered the most sacred, the mool mantra among the Vedic texts, and is taught for recitation among Hindus from an early age. Here, the inner spiritual meaning of the "Sun" becomes abundantly clear. The object of veneration is not that which provides us with light in the outside world but it is a principle that transcends the three planes of existence, the physical, the astral and the causal, and is the source of inner illumination.

This principle is referred to as Aum, a term whose three letters suggest the three phases of human experience: "A" referring to the waking state (jagrat), ".U" the dream state (swapna) and "M" the deep sleep state (sushupti).

The ultimate reality includes all three planes, and the three phases of human experience, yet goes beyond them. The silence that follows each recitation of the word Aum suggests the state of Turiya or Absolute Being, which is the indescribable source and end of everything. It is the Brahman the All-transcending One, whose prime attribute is effulgence, but who is in himself even beyond this effulgence. Hence the mantra in its original Rig Veda form has another line added to it, which is given out only to sanyasins and chosen disciples-Paro Raj-asal Savad Aum: He who transcends the effulgence is this Aum.

The Gayatri not only clarifies the routine implications of the references to the sun, abundant in the Vedas, but it also highlights another recurring theme in Hindu thought. Its wide imagery and popularity bring us to the question of mantras and their place in Indian religious practice.

The mantras or verbal formulae in Sanskrit verse or prose are classified into two types: those that are meant simply for recitation and need not be understood, and those that are divine invocations, whose import must be known in order to enable the devotee to keep his attention focused on the divine object.

The various mantras each have their individual benefits. There are those whose mastery or siddhi gives one contact with magical powers of a lower order (tamsic); there are others that bestow strength and courage and power (rajsic); and finally those whose sole object is sportual upliftment. Among the last as we have already seen, the Gayatri is the most venerated.

The mode of mantras, since time immemorial, stresses the spiritual importance of Sound. If the chanting of certain verbal formulae brings magical potential or assists spiritual advancement, then there must be latent in Sound itself an esoteric power. "This is why Vak Devi, the goddess of speech, was held in high esteem. Each word has its unique character and place, but of all words Aum is the most sacred. We have already examined some of its symbolic meanings. To these we may add still others. It is not only a term that connotes the qualities of the Absolute Brahman, but one that also denotes Brahman Himself. In the Rig Veda, we have:

Prajapati vai idam agref aseet
Tasya vag dvitya aseet,
Vag vai parmam Brahma.

("In the beginning was Prajapati, the Brahman, with whom was the Word and the Word was verily the Supreme Brahman.") This text remarkably parallels the opening of the Gospel according to St. John:

In the beginning was the Word, and the Word was with God and the Word was God.

Thus Aum becomes Brahman as manifesting Itself in the Word, and in the Taittriya Upanishad, It is referred to as the "Sheath of Brahman," as something which takes Its life from Brahman and contains Him. This aspect is made even clearer in the Sam Veda:

Brahman is at once Shabd and Ashabd both, And Brahman alone vibrates in space.

In other words, the Absolute One is not only inner effulgence but also beyond it, as suggested in the Gayatri. He is with the Word, the Shabd or Aum, yet beyond It. Both Sound and Light are in fact referred to as His prime manifestations. The Gayatri recommends that while concentrating on the Divine Word Aum, we fix our attention upon the inner Sun, while in the Chhandogya Upanishad, we are told that Naad, or the divine music, springs from the Universal Sun (of Brahmand), a secret that was given by Angris Rishi to Krishna, the darling son of Devki. It was this mystic insight to be found in the srutis, the scriptures revealed through inner hearing, that led to the development of what came to be called the Sphota-vada or the philosophy of the Word.

The teachers of this path preached that the Absolute was Wordless, imageless, indescribable and unconditioned. When He came into manifestation, He projected Himself as the Sphota or the Word, radiant with Light and vibrating with indescribable Music. The seeker wishing to transcend the relative plane to the Eternal and Unchanging must contact the Sphota or the Word Power through which he can rise to the Brahman who is beyond Shabd or Sphota. The Path of God-realization is certainly not easy. It is difficult to have access to, difficult to cognize, difficult to abide by and difficult to cross; yet it is the only possible Way, for one who is true to his Guru and His cause.

Such indeed are the truths that were taught and practiced by the forest sages of ancient India. But how much of them has survived since then? For the most part we find rituals such as the blowing of conches, ringing of bells, waving of lights, and the worship of the sun. These bear testimony to the mysteries within, but how few are conscious of their real significance?

In spite of Lord Krishna's powerful and lasting influence which brought the best of the Vedantic teachings to the heart of the common man, religion in India as elsewhere has tended to degenerate into mere caste and ceremony. The light and music outside are worshipped, but the flaming and sounding Word within, toward which they point, goes unheeded; "the light crieth in the darkness and the darkness comprehendeth it not."

The Hindu religion is difficult to describe. This is because it is so diverse. Some people who have studied the Holy Scriptures in depth may even say Hinduism is not really a religion; the true essence can be described as the "philosophy of life". Because Hinduism is based on philosophy it does not restrict the person to any one way of worshipping God, does not subscribe to one dogma, does not follow one set of religious rites or performances. In fact it presents to you *different philosophical concepts* of the same Eternal Truth.

"It may be with truth be asserted that no description of Hinduism can be exhaustive which does not touch on almost every religious and philosophical idea that the world has ever known. It is all-tolerant, all-comprehensive, all-compliant, all-absorbing. It has its spiritual and its material aspect; it's esoteric and exoteric; it's subjective and objective; it's rational and irrational. It has one side for the practical; another for the severely moral; another for the devotional and the imaginative; another for the philosophical and speculative."

Unity in Diversity

Hinduism is not a single religious faith system because it does not insist on a fixed set of doctrines. There are many sects and traditions within Hinduism. However despite this diversity, there is a unity among all this because the basic principles are based on the "eternal laws of nature"-SANATANA (eternal) DHARMA (laws of nature).

I.e. the best way to describe what Hindus practice is Hindu Dharma. Dharma is derived from the root "Dhru" meaning to uphold or support. Hindu Dharma provides a medium from which you are prevented from falling down and uplifted spiritually.

Hindu's Believe in One God

There is much confusion as to which God Hindu's believe in because of the many Gods Hindu's appear to worship. It is often seen as polytheism. Hindu's believe in one God. God can be formless; God has many aspects and these can be worshipped in giving God form. These "deities" (forms) can be human, male or female, or even animal form. Hindus point out there is one God Brahman. However to try and define God is to limit. Since God is limitless God cannot be defined. Hinduism allows people to find God in which ever way they wish.

"The apparent multiplication of gods is bewildering at the first glance, but you soon discover that they are the same GOD. There is always one uttermost God who defies personification. This makes Hinduism the most tolerant religion in the world, because its one transcendent God includes all possible gods. In fact Hinduism is so elastic and so subtle that the most profound Methodist, and crudest idolater, are equally at home with it."

George Bernard Shaw

This God "Brahman" is the same God worshipped by followers of all religions. Hindus believe the fundamentals of all religions are the same. The true essential nature of any religion is a path to the realization of Brahman. It does not matter if you call Brahman Allah, Jesus, Ram, Jehovah etc.

> *"If one religion is true, then all the others must also be true. Thus the Hindu faith is as yours as much as mine."*

Swami Vivekananda

What does Hinduism Have to Offer?

Most people search for religion when material things do not seem to satisfy themselves. The person begins to question "Where did I come from?" "Where will I go?" "Is life an empty meaningless accident?"

The Vedas (Holy Scriptures) reveal and explain Sanatana Dharma (the eternal laws of nature). They contain the answers to these questions. Hindu Dharma provides the philosophical basis and guidance for the person's quest for spirituality.

The Search for Happiness

The ancient Rishis (holy men) of India who compiled the Vedas and other Holy Scriptures described that everything we do is motivated by our search for one thing. Happiness.

Essentially there are two types of happiness. Impermanent (*e.g.* that which we receive from sense objects like a chocolate bar, or a cigarette) and permanent. Vedanta (the practice of the philosophy of the Vedas and other holy scriptures) provides guidance in achieving this "NIRVANA" (or moksha or self realization), this "PERMANENT BLISS", this "PERFECTION" which is achieved when the person realizes the ETERNAL TRUTH.

Hindu Values

Respect for Mother Earth
Respect for Mother
Respect for Father
Respect for Ancestors
Respect for Teachers (in other word respect everyone and everything)
Truth Satyaneva Jayate *"Truth alone triumphs"*
Mandikya Upanishad
Righteous living
Forgiveness and fearlessness
Honesty
Service to mankind
Non-violence
Love

The philosophy of Hinduism is extremely deep. I have tried to explain the basics but there is a lot more to it (reincarnation, the law of Karma, the different yogas etc).

"Each soul is potentially divine. The goal is to manifest this divinity within, by controlling nature, external and internal.

Do this either by work, or worship, or psychic control, or philosophy, by one or more of these and be free. This is the whole religion. Doctrines, or Dogmas, or rituals, or books, or temples, or forms are but secondary details."

The Five Sheaths

Kosas

Kosas or Koshas are five cases or sheaths which cover the Atman in Hinduism.

The five Koshas are:

1. Annamaya kosha, the food-body
2. Pranamaya kosha, the breath (*prana*)
3. Manomaya kosha, or "mind-sheath" (*Manas*)
4. Vijnanamaya kosha, that which discriminates, determines or wills (*Vijnana*)
5. Anandamaya kosha, or bliss (*Ananda*)

Corporal Bodies

In the consequence of acts performed in former states of being through the actions of the fivefold elements, corporal bodies (which become the dwelling place of pleasure and pain) are formed. The soul is wrapped in five investing sheaths and seems formed of these, and is darkened like crystals on coloured cloth. As winnowed rice is purified from husk, the soul burdened with its sheaths is purified by the force of meditation.

To remove his bondage the wise man should discriminate between the self and the non-self. By that alone he comes to know his own self and existence-through knowledge and bliss absolute, he becomes happy. He is free indeed who discriminates all sense-objects and the indwelling, unattached and inactive self as one separates a stalk from its enveloping sheath. Always merging every thing in it, he remains in a state of identity with that self.

Covered by the five sheaths, the material one and the rest, which are the products of its own power, the self ceases to appear, like the water of the tank by accumulation of sludge.

When all the five sheaths have been eliminated by the reasoning on Shruthi passages, what remains as the culminating point of the process is the witness, the knowledge absolute-the Atman. This self–effulgent Atman is distinct from the five sheaths, as witness of the three states, the real, the changeless, and the unattained everlasting bliss is to be realized by a wise man as his own self.

Five Sheaths

The five sheaths (pancha-kosas) are alluded to in the fourteen verse of the Atmabodha. It must be noted that the individualised soul, when separated from the Supreme soul, is regarded in the Vedanta as enclosed in a succession of cases (kosa) which envelope it and, as it were, folded one over the other, "like the coats of an onion". The five sheaths are said to cover the self. The true self or the Atman, is none of these, nor can its true nature be known as long as it is identified with them. All the five sheaths have been eliminated in the self of man. It appears pure, of the essence of everlasting and an alloyed bliss, indwelling, supreme, and self-effulgent.

Annamaya Kosa

The physical body is said to be made of food or matter. It is Annamaya. The Atman is described as enclosed in a series of sheaths. First there is the Annamaya, the food-body. This is the sheath of the physical self, named from the fact that it is nourished by the food. Further it says that the food identifies himself with a mass of skin, flesh, fat, bones, and filth, while the man of discrimination knows his own self, the only reality that there is, as distinct from the body. A materialist thinks he is the body, a religious student identifies himself with the mixture of body and Soul, while a Sage who has attained realization due to discrimination looks upon the eternal Atman as his self, and thinks "I am Brahman".

Pranamaya Kosa

There is another and subtle sheath, which is pranamaya, different from the body of food. Pranamaya means composed of prana. Prana is the vital principle, the force that vitalizes and holds together the body and the mind. It pervades the whole organism, its physical manifestation is the breath. As long as this vital principle exists in the organisms, life continues. The prana, with which we are all familiar, coupled with the five organs of action, forms the vital sheath, permeated by which the material sheath engages itself in all activities as if it were living. This is the sheath composed of breath and the other vital airs associated with the organs of action. In the Vivekachoodamani it is a modification of vayu, and like the air, it enters into and comes out of the body, and because it never knows in the least either its own weal and woo, nor those of others, being eternally dependent on the Self.

Manomaya Kosa

Manomaya means composed of manas or mind. It is called the manomaya or sheath composed of mere intellect, associated with the organs of action. This gives the individual soul its power of thought and judgement. For example, the *manomaya kosa*, or "mind-sheath", is said more truly to approximate to personhood than *annamaya kosa*, or the "food-sheath". The vital sheath knowledge together with the mind from the mental sheath the cause of the diversity of things such as *I* and *mine*. It is powerful and endowed with the faculty of creating categorical differentiation, such as names. It manifests itself as permeating the preceding.

Shankara uses the example of clouds that are brought in by the wind and again driven away by the same agency. Similarly, man's bondage is caused by the mind, and liberation, too, is caused by that alone. Attaining purity through a preponderance of discrimination and renunciation, the mind makes for liberation. Hence, the wise seeker after liberation must first strengthen these two. The mind (manas) along with the five sensory organs is said to constitute the manomaya kosa.

Vijnanamaya Kosa

Vijnanamaya means composed of vijnana, or intellect, and refers to the faculty which discriminates, determines or wills. Chattampi Swamikal defines vijnanamaya as the combination of intellect and the five sense organs. It is the sheath composed of more intellection, associated with the organs of perception. This gives the personal soul its first conception of individuality. The sheath of buddhi or vijnanamaya kosha together with the pranamaya kosa compose the body. Shankara holds that the buddhi, with its modifications and the organs of knowledge, form the vijnanamaya kosa or knowledge sheath, of the agent, having the characteristics which are the cause of man's transmigration.

This knowledge sheath, which seems to be followed by a reflection of the power of the chit, is a modification of prakriti. It is endowed by the function of knowledge, and it always wholly identifies itself with the body, organs etc. Owing to its connection with superimpositions, the supreme self, even though naturally, perfect and eternally unchanging, assumes the qualities of the superimpositions and appears to act just like the changeless fire assuming the modifications of the iron which it turns red-hot.

This knowledge sheath cannot be the supreme self for the following reasons

- It is subject to change.
- It is insentient.
- It is a limited thing.
- It is not constantly present.

Anandamaya Kosa

Anandamaya means composed of ananda (bliss), and its sheath refers to the ego. In the Upanishads the sheath is known also as the *causal body*. In deep sleep, when the mind and senses cease functioning, there still stands the causal body between the finite world and the blissful self. As the fifth sheath is nearest of all to the blissful self, its name in the Upanishads is *anandamaya*. Anandamaya, or that which is composed of Supreme bliss, is also named, although not admitted, by all. It is regarded as the innermost of all, and therefore, when the five sheaths are enumerated, it is to be placed before the vijnanamaya. The blissful sheath has its fullest play during deep sleep, while in the dreaming and wakeful states, it has only a partial manifestation. The blissful sheath (anandamaya kosha) is that modification nescience which manifests itself by catching a reflection of the Atman which is bliss absolute, whose attributes are pleasure and rest, and which comes into view when some object agreeable to oneself presents itself. It makes itself spontaneously felt by the fortunate during the fruition of their virtuous deeds, from which every corporeal being derives great joy without effort.

The Five Sheaths

This chapter is based on the teachings of Vedanta as embodied in the major Hindu scriptures—the Upanishads, the Bhagavad Gita, the Brahma Sutras, and others—and interpreted by Shankaracharya, the exponent of non-dualism. According to non-dualism, the Ultimate Reality of everything is Brahman, the non-dual pure consciousness, and It alone exists. The universe of beings and things is merely an appearance of Brahman in time and space.

The individual soul and Brahman are absolutely non-different. The root cause of all bondage is the soul's ignorance of its true nature. Liberation is union with Brahman attained through Self-knowledge. The two other interpretations of Vedanta are qualified non-dualism and dualism. The chief exponent of qualified non-dualism is Ramanuja and of dualism, Madhva. Qualified non-dualism maintains that Brahman, though non-dual pure consciousness, transforms Itself into God, the universe, and the world of souls, and that the transformation is real.

God is the whole and the individual soul is the part. The bondage of the individual soul is due to its alienation from God, and liberation is communion with God. Dualism believes that God is a personal being who creates the universe and the world of souls. The creation is real and the created beings and things are different from and dependent on God. The bondage of the soul is due to its forgetfulness of God, and liberation is communion with God.

Background of the Faith

Hinduism, considered the oldest religion of the world, is today practiced by over 500 million people in India and other countries. The word "Hindu" is a distorted form of "Sindhu," the Sanskrit word for Indus, the river that flows into the Arabian Sea. This mispronunciation is attributed to the Persians who invaded India at the end of the sixth century BC. Later, during the Greek invasion of India (326 BC), the Greeks described the river Sindhu as "Indos," which was sometime afterward changed to "Indus." Since then, the country east of the river Indus has come to be known as India, its people as Indians, and their religion as Hinduism. The original name of Hinduism is *Sanatana Dharma*, which means "eternal religion." It was also known as *Vaidika Dharma*, or the "religion of the Vedas." The European traders and Christian missionaries who came to India at the beginning of the seventeenth century signified Hinduism as "Brahmanism."

A federation of many systems of thought, Hinduism is based not on any fixed sets of creeds and dogmas but on certain eternal principles. It was not founded by any historical personality. Many prophets, saints, mystics, and philosophers, both ancient and modern, have contributed to its growth, development, and perpetuation. The great teachers of Hinduism are Sri Rama (of the Ramayana), Sri Krishna (of the Bhagavad Gita), Sri Shankaracharya (AD 788-820), Sri Ramanuja (AD 1017-1137), Sri Madhva (AD 1199-1276), Sri Chaitanya (AD 1485-1533), and in modern times, Sri Ramakrishna (AD 1836-1886), and Swami Vivekananda (AD 1863-1902).

Their lives demonstrate the validity of the spiritual teachings of Hinduism. Hinduism derives its authority primarily from the four Vedas: the Rig-Veda, the Yajur-Veda, the Sama-Veda, and the Atharva-Veda. Each Veda consists of four parts: the *mantras*, or hymns in praise of Vedic deities; the *brahmanas*, or the section dealing with rituals and ceremonies; the *aranyakas*, or philosophical interpretation of the rituals; and the Upanishads, or the concluding portions of the Vedas (known as Vedanta), which describe the profound spiritual truths. Hinduism derives its authority secondarily from another group of scriptures-the Ramayana, the Mahabharata, the Bhagavad Gita, the Puranas, and others. Of the two groups of scriptures, the Vedas along with the Upanishads are known as *srutis*, while the others are called *smritis*. *Sruti* is revelation and *smriti* is tradition.

The Ultimate Reality

Hindu scriptures describe ultimate reality as Brahman. Brahman is non-dual pure consciousness, indivisible, incorporeal, infinite, and all-pervading like the sky. Brahman is of the nature of existence-knowledge-bliss-absolute-the ground of all existence, basis of all awareness, and source of all bliss. It is the reality of all realities, the soul of all souls, one without a second, the constant witness of the changing phenomena of the universe.

From the absolute point of view, Brahman alone exists. Brahman has two aspects: transcendent and immanent. In Its transcendent aspect, Brahman is devoid of name and form, sex and attributes. But in Its immanent aspect, Brahman is endowed with them. The Upanishads designate the transcendent Brahman by the word "It" and the immanent Brahman by the word "He." Through Its inscrutable power called *maya*, the transcendent Brahman appears to be conditioned by time and space and to manifest itself as personal God, the creator, preserver, and destroyer of the universe. The Upanishads describe God as the supreme person:

His hands and feet are everywhere; His eyes, heads, and faces are everywhere; His ears are everywhere; He exists compassing all. [i] The heavens are His head; the sun and moon, His eyes; the quarters, His ears; the revealed Vedas, His speech; the wind is His breath; the universe, His heart. From His feet is produced the earth. He is, indeed, the inner Self of all beings. [ii] The various Godheads of Hinduism, such as Shiva, Vishnu, Brahma, Kali, and Durga, are but different facets of Brahman. The supreme Brahman assumes various forms for the fulfilment of the individual spiritual seekers. All concepts and forms of God, according to Hinduism, are what we think of Him and not what He is to Himself. Again, various seekers of God, depending upon their advancement, perceive God differently. For example, to the beginner God appears as an extra-cosmic creator; to the more advanced seeker as inner controller; and to the perfect knower of God, God is everywhere and in everything. Still another manifestation of the conditioned Brahman is the incarnation of God-God's taking human form. According to Hinduism, God incarnates Himself to fulfil the needs of the universe, whenever and wherever such a need arises. In the Bhagavad Gita Sri Krishna says:

> *Whenever there is a decline of* dharma *(righteousness), O Bharata, and a rise of* adharma *(unrighteousness), I incarnate myself. For the protection of the good, for the destruction of the wicked, and for the establishment of* dharma, *I am born in every age. Thus, according to Hinduism, the supreme Godhead is both formless and endowed with many forms.*

The Universe

Hinduism maintains that the universe is beginning less and endless. It subscribes to the theory of manifestation and non-manifestation of the universe, of evolution of cosmic energy into names and forms and its involution. The Vedas describe this process as the out-breathing and in-breathing of Brahman. The Upanishads say that just as the hair and nails grow on a living person, as the threads come out of a spider, as sparks fly from a blazing fire, as melodies issue from a flute, or as waves rise on the ocean, so also does the universe come forth from Brahman. Brahman is both the material and the efficient cause of the universe.

This manifestation of Brahman as the manifold universe is not real but apparent. Through its inscrutable power of *maya*, Brahman appears as the world of matter and souls, and as endowed with the activities of creation, preservation, and dissolution. Maya veils the ultimate reality and in its place projects various appearances. *Maya* is change and relativity. It is neither real nor unreal nor both. If the world of *maya* were real, then it could never be changed. On the other hand, it cannot be unreal because the sufferings of life are felt tangibly.

As long as it is not known, *maya* is delusive; but when known, *maya* is nothing but Brahman. *Maya* is comprised of the three gunas, or qualities: sattva, rajas, and tamas. Sattva is balance or equilibrium; rajas is restlessness or imbalance; tamas is inertia or darkness.

The three gunas are present in varying degrees in all objects, gross or subtle, including the body-mind complex of an individual. For example, when sattva prevails in an individual, the light of knowledge begins to shine through his body and mind. When rajas prevails, he is stirred by unrest. And when tamas prevails, he is taken over by inertia. When the universe is in a period of non-manifestation, the three gunas remain in a state of non-differentiation, or equilibrium. Manifestation begins when the equilibrium of the gunas is disturbed.

According to Hinduism, the process of manifestation and non-manifestation of the universe follows a cyclical pattern. In each cycle there is a recurrence of the same material phenomena, and the same recurrences continue throughout eternity. No energy can be annihilated; it goes on changing until it returns to the source. Nature presents both movements-from the subtle to the gross and back from gross to subtle. Evolution presupposes involution.

Only that which was involved before can be evolved afterwards. Evolution of the physical universe follows a graduated process. The first element to evolve at the beginning of a cycle is akasa, or the ether, in its subtle form. Then gradually evolve four other elements: air, fire, water, and earth. In the beginning, the five elements remain unmixed. Then, through their various combinations, the elements take their gross forms. From out of the basic gross and subtle elements are produced all objects, gross and subtle, including the body-mind complex of all living creatures.

According to the Puranas, each world period is divided into four ages, or yugas: Satya, Treta, Dwapara, and Kali. The Satya yuga abounds in virtue, with vice being practically non-existent. But with each succeeding age, virtue gradually diminishes and vice increases, making the age of Kali the reverse of Satya. The approximate duration of each yuga has been described as: Satya yoga, 1,728,000 years; Treta, 1,296,000 years; Dwapara, 864,000 years; and Kali, 432,000 years.

These four yugas, rotating a thousand times, make one day of Brahma, the creator, and an equal number of years, one night. Thirty such days and nights make one month of Brahma, and twelve months make one year. After living for a hundred such years, Brahma dies. Brahma, too, like all other entities of the phenomenal universe, has a limited life span, although this life span seems nearly endless from the viewpoint of human calculations.

The Human Individual

According to Hinduism, man is essentially a soul that uses its body and mind as instruments to gain experience. What is the nature of the soul? Hinduism maintains that the macrocosm and the microcosm are built on the same plan, and that Brahman is the soul of both. As the soul of man, Brahman is known as Paramatman. The Upanishads speak of the two souls of man dwelling, as it were, side by side, within him: the real soul (Paramatman) and the apparent soul (jivatman). The real soul is the witness consciousness, serene and detached. The apparent soul is the embodied soul, the experiencer of birth and death, and is ever in quest of freedom and eternal life.

The apparent soul is the ego self—the reflection of the real soul. The real soul has been described as Self and the apparent soul as non-Self. Hinduism analyzes man in terms of three bodies, five sheaths, and three states. It says that a human individual has three bodies: physical body, subtle body, and causal body. The physical body is produced out of the gross forms of the five basic elements (ether, air, water, fire, and earth), and is subject to a sixfold change: birth, subsistence, growth, maturity, decay, and death. At death the physical body perishes and its five constituent elements are dissolved. The subtle body is made of the subtle forms of the five basic elements that produced the physical body.

It is the receptacle of thoughts and memories and continues to exist after death, serving as the vehicle of transmigration.

A human individual enters this world with a bundle of thoughts in the form of his mind, and he also exits with a bundle of thoughts, some old and some new. The causal body, characterized by ego sense only, is finer than the subtle body. All three bodies are for the fulfilment of desires, gross and subtle. The soul is different from these three bodies.

Hindu scriptures further describe the body-mind complex of man as consisting of five sheaths, or layers: the physical sheath, the sheath of prana (the vital air), the sheath of mind, the sheath of intellect, and the sheath of bliss. These sheaths are located one inside the other like the segments of a collapsible telescope, with the sheath of the physical body being the outermost and the sheath of bliss being the innermost. The sheath of the physical body is dependent on food for its sustenance and lasts as long as it can absorb nourishment. The sheath of the vital air is the manifestation of the universal vital energy.

It animates the gross body, making it inhale and exhale, move about, take in nourishment, excrete, and reproduce. The sheath of the mind is the seedbed of all desires. It is changeful, characterized by pain and pleasure, and has a beginning and an end. The sheath of the intellect is the seat of I-consciousness. Though material and insentient by nature, it appears intelligent because it reflects the light of the Self. It is the cause of embodiment. Finer than the sheath of the intellect is the sheath of bliss, the main features of which are pleasure and rest. It, too, is material and subject to change. The five sheaths are the five layers of embodiment and they veil the light of the Self.

The Upanishads mention that man experiences three states of existence-waking, dream, and deep sleep-and his Self within, the experiencer of the three states, is different from them. Analysis of all three states reveals the true nature of the Self. In the waking state man remains identified with his physical body, in the dream state with his subtle body, and in deep sleep with his causal body.

The Atman, or Self, is the monitoring consciousness of all three states and is the basis of their unity. Hinduism contends that conclusions based only on an analysis of the waking state are incomplete and cannot reveal the real nature of man. In this sense, Hinduism considers the conclusions of physical science as inadequate although not incorrect.

The Five Sheaths

1, 2. The Guru said: This Annamaya sheath or good sheath is made up of the five elements. It has a beginning and an end. It is inert and full of parts. It is an effect of the five elements. It is full of impurities. Therefore you are not this physical body or the Annamaya sheath. You are the witness of this body. Understand, therefore, "I am not the body. I am Brahman."

The physical body is the grossest form of thought. The food consumed by the parents is converted into Sukla (semen) in men and Sonita in women and by the combination of these the physical body is formed. After birth, the body grows by suckling the milk which is only a transformation of the food consumed by the mother. The body is further developed by taking food. It gets dissolved in earth which is another form of food. The body is itself a food for other creatures. Hence it is called the food sheath, the material body or the earthly encagement of the soul. The food sheath is an object of perception. The Atman is the cogniser and the body is the cognised. Hence the Self is different from the body. In dream and deep sleep there is no consciousness of the body.

The five elements constitute the physical body. These modifications of Maya are not the Truth, the body and its Dharmas, size, form, birth and death are not actual modifications of the Self. Varnashrama, name and class differ in different births. They are mere accidental attributes of the body. There is no physical body either before birth or after death. Hence it is non-eternal.

Existence, birth, growth, modification, decay and death are the six Vikaras of the physical body. Just as the ether in a pot is not affected in any way by the destruction of the pot, so also the Atman is not at all affected by the destruction of the body or the Annamaya Kosha. Atman is unattached. Ether is subtle, but the Atman is still subtler. Atman is formless, changeless, birthless, deathless, free from old age. It is neither born nor is killed. Hence one should meditate on this Atman or Brahman.

3, 4. The Pranamaya Kosha or the vital sheath is a product of Rajoguna. It also has a beginning and an end. It is inert. It is an effect. Therefore you are not the Pranamaya Kosha. You are the witness of this sheath. Understand, therefore, "I am not the Pranamaya Kosha. I am Brahman."

The Pranamaya Kosha consists of the five Pranas and five Karma-Indriyas or organs of action. Though the Prana is waking when one is sleeping, it does not invite a friend and entertain him; it cannot stop a thief who tries to remove the articles in a house. Therefore it is insentient. The Self is a mass of Intelligence. It is Chaitanya-Swarupa. It is entirely different from the Prana. The Self is the knower, seer and witness of this sheath.

Prana is only the active working of the mind. A pure-hearted man breathes rhythmically. The breath of an evil-minded person is disturbed. When the mind is controlled the Prana is automatically controlled. The Vedantic aspirant does not practise Pranayama, because his breath is automatically regulated and Kumbhaka naturally follows when the mental Kumbhaka or concentration and meditation are practised. The Pranas are the Rajasic manifestations of the dynamic mental force which with their ups and downs maintain the balance of individual existence even as the bicycle is kept in balance when its wheels are vigorously turning. When there is a break of this movement, the bicycle falls down and when the Prana is inhibited the individualising mind together with the ego breaks down and dies.

Hence there should be no identification with the Pranamaya Kosha and the aspirant should assert the Self-existent Atman distinct from it.

5, 6. The Manomaya Kosha or the mental sheath is a product of Sattwa Guna. It also has a beginning and an end. It is inert. It is an effect. Therefore you are not the

Manomaya Kosha. You are the witness of this sheath. Understand, therefore, "I am not Manomaya Kosha. I am Brahman."

The Manomaya Kosha consists of the mind and the five Jnana Indriyas. It is a means of enjoying pleasure and pain. The mind causes egoism in the body and "mine"-ness in house, sons, wife, wealth, etc., and passes outside through the avenues or channels of these five Indriyas. It is the internal instrument for gaining the experiences and knowledge of this world. Mind is associated with the Vrittis or waves of lust, anger, etc., and is a terrible objectifying agent. Mind is a Vikari, it constantly changes itself.

The Self is a witness of the Manomaya Kosha. The Self is Nirvikari. The mind is not the Self. The Self is the Atman or Brahman, unblemished, eternal and changeless, and one should meditate on it as such.

7, 8. The Vijnanamaya Kosha or this Buddhi sheath is a product of Sattwa Guna. It has also a beginning and an end. It is inert. It is an effect. Therefore you are not the Vijnanamaya Kosha. You are witness of this sheath. Understand, therefore, "I am not the Vijnanamaya Kosha. I am Brahman."

The Vijnanamaya Kosha consists of the intellect in conjunction with the five organs of knowledge or the Jnana-Indriyas. During sleep it gets involution or Laya along with Chidabhasa or the reflection of Pure Consciousness. During waking state it is the doer. It is an effect like a jar and is inanimate. It shines in borrowed feathers. It borrows its light temporarily from its source, just as the moon borrows its light from the sun. It is not the eternal Self.

The Pranamaya, Manomaya and the Vijnanamaya Koshas constitute the subtle body. The subtle body is composed of the five unquintuplicated elements. There is neither breathing nor talking, neither seeing nor hearing in the dead body. There is also no warmth. The self-cognitions such as "I speak; I hear; I am hungry; I am thirsty;" and the like appear distinctly in the subtle body. The subtle body operates in the waking and the dreaming states. Ghosts and apparitions are the manifestations of the subtle body only.

The ego is hidden in the intellect and the memory (Chitta) is hidden in the mind. The subtle body thus, contains nineteen principles or Tattwas. It is also called the "Puri-Ashtaka" or the eightfold city. The five organs of sense, the five organs of action, the five vital breaths, the five subtle primary elements, the fourfold Antahkarana, ignorance, desire and action are the eightfold city of the subtle body.

The physical body is only an instrument in the hands of the subtle body. When the subtle body is disciplined through Pranayama, abstraction and concentration, the physical body also becomes very healthy and strong. Whatever the subtle body is, that the physical body also becomes. The mind which is the ruler of the subtle body gets fattened by worldly affections, by avarice for wealth, by the acquirement of women and gold and by attachment to the external fleeting forms of beauties. The mind is thinned out by eradication of the Vasanas and egoism.

The subtle body is the distracted expression of the self through Avidya, the causal sheath. Therefore it is not the Truth. Truth is Brahman and all else is false. One should meditate that he is not the subtle body and that he is the self-effulgent Atman.

9, 10. The Anandamaya Kosha or this bliss sheath is Avidya or ignorance, a modification of Prakriti. It is the effect of past deeds. It is endowed with changing attributes. It is Jada or insentient. Therefore you are not the Anandamaya Kosha. You are the witness of this sheath. Understand, therefore, "I am not the Anandamaya sheath. I am Brahman."

The Anandamaya Kosha is made of Mula-Ajnana. It is the Karana Sarira or the causal body which is the substratum of all other sheaths which are external to it. Its three attributes or Dharmas are Priya, Moda and Pramoda, affection, delight and intense happiness. It is the indescribable beginningless Avidya, the nescience of the Atma, and is composed of Malina

Sattwa. It is inanimate, beginningless, but has an end in Atma-Jnana.

The ignorance of the real nature of the Self constitutes this causal body or seed-body. It contains the potentialities or the seeds for the subtle and gross bodies. It projects the appearance of the whole universe through the subtle sheath. It is the food of ignorance for the hungry ego.

The mind has come out of this ignorance and gets involved in it during deep sleep. In the sleeping state there is a vigorous functioning of this ignorance in which everything is lost as in pitch darkness. The Karana Sarira screens the Satchidananda Brahman.

He who knows the ignorance or the negation of the existence of the Atman and the denial of its appearance is the true Self, the Atman. He who knows the effects of ignorance, such as "I am a man, I am the doer and enjoyer, I am happy, I am miserable," is the witness and the Atman. Hence in reality the Self is the seer, knower and the witness of the causal body or the ignorance. The Self is the Knowledge and the Light itself.

As the light that enlightens the jar is different from it, so is the Self different from the bodies witnessed by it. Therefore the Self is Consciousness itself and not the bodies.

The aspirant should endeavour to rise above the five Koshas to realise the identity with Pure Consciousness. Just as one draws out the thin stalk from the Munja grass by stripping off its upper layers one by one, so also one should take out the innermost essence of the Atman from all objects of perception, *i.e.* the five Koshas, by the "neti, neti" doctrine of negating unreality.

Just as butter is removed from milk by churning the mixture of curd, so also the butter of the Atman should be taken from the mixture of the five Koshas by the churning of constant meditation on the Immortal Brahman which fictitiously appears as the sheaths, the world, etc. When the identification with the sheaths ceases, the self realises the Infinite Being and gets liberated beyond death.

Symbols

Part-1: Important Hindu Symbols Explained!

Hinduism employs the art of symbolism with amazing effect. No religion is so replete with symbols as this ancient religion. And all Hindus are touched by this all-pervasive symbolism all through the life in some way or the other. Basic Hindu symbolism is enunciated in the *Dharmashastras*, but much of it developed with the evolution of his unique 'way of life'. On the surface, many Hindu symbols may seem to be absurd or even dumb, but discovering the deeper meaning of the such symbolism is sheer joy!

Om or Aum

As the cross is to Christians, the Om is to Hindus. It is made up of three Sanskrit letters, *aa*, *au* and *ma* which, when combined, make the sound *Aum* or *Om*. The most important symbol in Hinduism, it occurs in every prayer and invocation to most deities begins with it. As the symbol of piety, Om is often found at the head of letters, pendants, enshrined in every Hindu temple and family shrines.

This symbol is actually a sacred syllable representing the Brahman or the Absolute-the source of all existence. Brahman, in itself, is incomprehensible so a symbol becomes mandatory to help us realize the Unknowable. The syllable Om occurs even in English words having a similar meaning, for instance, 'omniscience', 'omnipotent', 'omnipresent'. Thus Om is also used to signify divinity and authority. Its similarity with the Latin 'M' as also to the Greek letter 'Omega' is discernable. Even the word 'Amen' used by Christians to conclude a prayer seems to be akin to Om.

Swastika

Second in importance only to the Om, the Swastika, a symbol which look like the Nazi emblem, holds a great religious significance for the Hindus. Swastika is not a syllable or a letter, but a pictorial character in the shape of a cross with branches bent at right angles and facing in a clockwise direction. A must for all religious celebrations and festivals, Swastika symbolizes the eternal nature of the Brahman, for it points in all directions, thus representing the omnipresence of the Absolute.

The term 'Swastika' is believed to be a fusion of the two Sanskrit words 'Su' (good) and 'Asati' (to exist), which when combined means 'May Good Prevail'. Historians say Swastika could have represented a real structure and that in ancient times forts were built for defense reasons in a shape closely resembling the Swastika. For its protective power this shape began to be sanctified.

The Saffron Colour

If there is any colour that can symbolize all aspects of Hinduism, it's saffron-the colour of Agni or fire, which symbolizes the Supreme Being. The saffron colour, also auspicious to the Sikhs, the Buddhists and the Jains, seems to have obtained religious significance much before these religions came into being. Fire worship had its origin in the Vedic age. The foremost hymn in the Rig Veda glorifies fire: "*Agnimile purohitam yagnasya devam rtvijam, hotaram ratna dhatamam.*" When sages moved from one *ashram* to another, it was customary to carry fire along. The inconvenience to carry a burning substance over long distances may have given rise to the symbol of a saffron flag. Triangular and often forked saffron flags are seen fluttering atop most Sikh and Hindu temples. While Sikhs regard it as a militant colour, Buddhist monks and Hindu saints wear robes of this colour as a mark of renunciation of material life.

Vedic rituals, like the 'Yagna' and 'Puja', as said Rishi Aurobindo, are "attempts to fulfil the purpose of creation and elevate the status of man to that of a godhead or a cosmic man". Puja is essentially a ritual suggestive of symbolic offering of our lives and activities to God.

Symbolic Significance of Puja Items

Every object associated with the ritual of Puja or worship is symbolically significant. The statue or image of the deity, which is called 'Vigraha' (Sanskrit: 'vi'+ 'graha') means something that is devoid of the ill effects of the planets or 'grahas'. The flower that we offer to the deity stands for the good that has blossomed in us. The fruits offered symbolize our detachment, self-sacrifice and surrender, and the incense we burn collectively stands for the desires we have for various things in life. The lamp we light represents the light in us, that is the soul, which we offer to the Absolute. The vermilion or red powder stands for our emotions.

The Lotus

The holiest of flowers for Hindus, the beautiful lotus is symbolic of the true soul of an individual. It represents the being, which lives in turbid waters yet rises up and blossoms to the point of enlightenment. Mythologically speaking, lotus is also a symbol of creation, since Brahma, the creator came forth from the lotus that blooms from the navel of Vishnu. It is also famous as the symbol of BJP-the Hindu Right-wing political party of India, the familiar lotus position in meditation and yoga, and as the national India and Bangladesh.

The Purnakumbha

An earthen pot or pitcher-called 'Purnakumbha'-full of water, and with fresh mango leaves and a coconut atop it, is generally placed as the chief deity or by the side of the deity before starting a Puja. Purnakumbha literally means a 'full pitcher' (Sanskrit: 'purna' = full, 'kumbha' = pot). The pot symbolizes mother earth, the water life-giver, the leaves life and the coconut divine consciousness. Commonly used during almost all religious rites, the pitcher also stands for goddess Lakshmi.

Fruits & Leaves

The water in the Purnakumbha and the coconut have been objects of worship since the Vedic age. The coconut (Sanskrit: Sriphala = God's fruit) alone is also used to symbolize 'God'. While worshipping any deity, a coconut is almost always offered along with flowers and incense sticks. Other natural objects that symbolize divinity are the betel leaf, the areca-nut or betel-nut, banyan leaf and the leaf of 'bel' or vilva tree.

Naivedya or Prasad

It is our ignorance ('avidya') which we offer to the deity in a Puja. The food symbolically stands for our ignorant consciousness, which we place before god for spiritual enlightenment. After he suffuses it with knowledge and light and breathes a new life into our bodies, it makes us divine. When we share the prasad with others, we share the knowledge we thus gained with fellowbeings.

The Vedic deities symbolize the forces in nature as well as inside human beings. While discussing the symbolic significance of Vedic deities in his *The Secret of the Vedas*, Shri Aurobindo says that the gods, goddesses and demons mentioned in the Vedas represent various cosmic powers, on one hand and man's virtues and vices on the other.

Why Worship Idol?

Idol worship and rituals are at the heart of Hinduism have great religious and philosophical significance. All Hindu deities are themselves symbols of the abstract Absolute, and point to a particular aspect of the Brahman. The Hindu Trinity is represented by three Godheads: Brahma-the creator, Vishnu-the protector and Shiva-the destroyer.

Why Worship Different Deities?

Unlike the followers of any other religion, Hindus enjoy the freedom of worshipping their personally chosen icon to offer their prayers to the indefinable Brahman. Each deity in Hinduism controls a particular energy. These energies, present in man as wild forces must be controlled and canalized fruitfully to infuse a divine consciousness in him.

For this, man has to gain the goodwill of different gods who stir up his consciousness accordingly to help him master the different forces of nature. In a person's path of spiritual progress, he or she needs to develop the various attributes of these godheads in him or her to attain all-round spiritual perfection.

Symbolism of Gods & Goddesses

Each Hindu God and Goddess has many characteristics, like dress, 'vehicle', weapons, etc., that are themselves symbols of the deity's power. Brahma holds the Vedas in his hands, which signifies that he has the supreme command over creative and religious knowledge. Vishnu holds a conch which stands for the five elements and eternity; a discus, which is the symbol of the mind; a bow that symbolizes power and a lotus which is the symbol of the cosmos. Shiva's trident represents the three *gunas*. Similarly Krishna's flute symbolizes divine music.

Many deities can be recognized by the symbols associated with them. Shiva is often symbolized by the 'linga' or the three horizontal lines on his forehead. In the same way Krishna can be identified by the peacock feather he wears in his head and also by the prong-like mark on his forehead.

Vehicles of the Gods

Each deity has a particular vehicle on which he or she travels. These vehicle, which are either animals or birds, represent the various forces that he or she rides. Goddess Saraswati's vehicle, the graceful and beautiful peacock denotes that she is the controller of the pursuit of performing arts. Vishnu sits on the primal serpent, which represents the desire of consciousness in humankind.

Shiva rides the Nandi bull, which stands for the brute and blind power, as well as the unbridled sexual energy in man-the qualities only he can help us control. His consort Parvati, Durga or Kali rides on a lion, which symbolizes mercilessness, anger and pride-vices she can help her devotees check. Ganesha's carrier, a mouse represents the timidity and nervousness that overwhelm us at the onset of any new venture-feelings that can be overcome by the blessings of Ganesha.

3

The Purpose of Life of a Hindu

To fulfil our duty in the divine scheme we must try to understand not only that scheme as a whole, but the special part that man is intended to play in it.

The divine outbreathing reached its deepest immersion in matter in the mineral kingdom, but it reaches its ultimate point of differentiation not at the lowest level of materiality, but at the entrance into the human kingdom on the upward arc of evolution. We have thus to realize three stages in the course of this evolution.

(a) The downward arc in which the tendency is towards differentiation and also towards greater materiality. In this stage spirit is involving itself in matter, in order that it may learn to receive impressions through it.

(b) The earlier part of the upward arc, in which the tendency is still towards greater differentiation, but at the same time towards spiritualization and escape from materiality. In this stage the spirit is learning to dominate matter and to see it as an expression of itself.

(c) The later part of the upward arc, when differentiation has been finally accomplished, and the tendency is towards unity as well as towards greater spirituality. In this stage the spirit, having learnt perfectly how to receive impression through matter and how to express itself through it, and having awakened its dormant powers, learns to use these powers rightly in the service of the Deity.

The object of the whole previous evolution has been to produce the ego as a manifestation of the Monad. Then the ego in its turn evolves by putting itself down into a succession of personalities. Men who do not understand this look upon the personality as the self, and consequently live for it alone, and try to regulate their lives for what appears to be its temporary advantage.

The man who understands realizes that the only important thing is the life of the ego, and that its progress is the object for which the temporary personality must be used. Therefore when he has to decide between two possible courses he thinks not, as the ordinary man might: "Which will bring the greater pleasure and profit to me as a personality?" but "Which will bring greater progress to me as an ego?" Experience soon teaches him that nothing can ever be really good for him, or for anyone, which is not good for all, and so presently he learns to forget himself altogether, and to ask only what will be best for humanity as a whole.

Clearly then at this stage of evolution whatever tends to unity, whatever tends to spirituality, is in accord with the plan of the Deity for us, and is therefore right for us, while whatever tends to separateness or to materiality is equally certainly wrong for us. There are thoughts and emotions which tend to unity, such as love, sympathy, reverence, benevolence; there are others which tend to disunion, such as hatred, jealousy, envy, pride, cruelty, fear. Obviously the former group are for us the right, the latter group are for us the wrong. In all these thoughts and feelings which are clearly wrong, we recognize one dominant note, the thought of self; while in all those which are clearly right we recognize that the thought is turned toward others, and that the personal self is forgotten.

Wherefore we see that selfishness is the one great wrong, and that perfect unselfishness is the crown of all virtue. This gives us at once a rule of life. The man who wishes intelligently to co-operate with the Divine Will must lay aside all thought of the advantage or pleasure of the personal self, and must devote himself exclusively to carrying out that Will by working for the welfare and happiness of others.

This is a high ideal, and difficult of attainment, because there lies behind us such a long history of selfishness. Most of us are as yet far from the purely altruistic attitude; how are we to go to work to attain it, lacking as we do the necessary intensity in so many of the good qualities, and possessing so many which are undesirable?

Here comes into operation the great law of cause and effect to which I have already referred. Just as we can confidently appeal to the laws of Nature in the physical world, so may we also appeal to these laws of the higher world. If we find evil qualities within us, they have grown up by slow degrees through ignorance and through self-indulgence. Now that the ignorance is dispelled by knowledge, now that in consequence we recognize the quality as an evil, the method of getting rid of it lies obviously before us.

For each of these vices there is a contrary virtue; if we find one of them rearing its head within us, let us immediately determine deliberately to develop within ourselves the contrary virtue. If a man realizes that in the past he has been selfish, that means that he has set up within himself the habit of thinking of himself first and pleasing himself, of consulting his own convenience or his pleasure without due thought of the effect upon others; let him set to work purposefully to form the exactly opposite habit, to make a practice before doing anything of thinking how it will affect all those around him; let him set himself habitually to please others, even though it be at the cost of trouble or privation for himself. This also in time will become a habit, and by developing it he will have killed out the other.

If a man finds himself full of suspicion, ready always to assign evil motives to the actions of those about him, let him set himself steadily to cultivate trust in his fellows, to give them credit always for the highest possible motives. It may be said that a man who does this will lay himself open to be deceived, and that in many cases his confidence will be misplaced. That is a small matter; it is far better for him that he should sometimes be deceived as a result of his trust in his fellows than that he should save himself from such deception by maintaining a constant attitude of suspicion. Besides, confidence begets faithfulness. A man who is trusted will generally prove himself worthy of the trust, whereas a man who is suspected is likely presently to justify the suspicion.

If a man finds in himself the tendency towards avarice, let him go out of his way to be especially generous; if he finds himself irritable, let him definitely train himself in calmness; if he finds himself devoured by curiosity, let him deliberately refuse again and again to gratify that curiosity; if he is liable to fits of depression, let him persistently cultivate cheerfulness, even under the most adverse circumstances.

In every case the existence of an evil quality in the personality means a lack of the corresponding good quality in the ego. The shortest way to get rid of that evil and to prevent its reappearance is to fill the gap in the ego, and the good quality which is thus developed will show itself as an integral part of the man's character through all his future lives. An ego cannot be evil, but he can be imperfect.

The qualities which he develops cannot be other than good qualities, and when they are well defined they show themselves in each of all his numerous personalities, and consequently those personalities can never be guilty of the vices opposite to these qualities; but where there is a gap in the ego, where there is a quality undeveloped, there is nothing inherent in the personality to check the growth of the opposite vice; and since others in the world about him already possess that vice, and man is an imitative animal, it is quite probable that it will speedily manifest itself in him.

This vice, however, belongs to the vehicles only and not to the man inside. In these vehicles its repetition may set up a momentum which is hard to conquer; but if the ego bestirs himself to create in himself the opposite virtue, the vice is cut off at its root, and can no longer exist—neither in this life nor in all the lives that are to come.

A man who is trying to evolve these qualities in him-self will find certain obstacles in his way—obstacles which he must learn to surmount. One of these is the critical spirit of the age—

the disposition to find fault with a thing, to belittle everything, to look for faults in everything and everyone. The exact opposite of this is what is needed for progress. He who wishes to move rapidly along the path of evolution must learn to see good in everything—to see the latent Deity in everything and in everyone. Only so can he help those other people—only so can he get the best out of those other things.

Another obstacle is the lack of perseverance. We tend in these days to be impatient; if we try any plan we expect immediate results from it, and if we do not get them, we give up that plan and try something else. That is not the way to make progress in occultism. The effort which we are making is to compress into one or two lives the evolution which would naturally take perhaps a hundred lives.

That is not the sort of undertaking in which immediate results are to be expected. We attempt to uproot an evil habit, and we find it hard work; why? Because we have indulged in that practice for, perhaps, twenty thousand years; one cannot shake off the custom of twenty thousand years in a day or two. We have allowed that habit to gain an enormous momentum, and before we can set up a force in the opposite direction we have to overcome that momentum.

That cannot be done in a moment, but it is absolutely certain that it *will* be done eventually, if we persevere, because the momentum, however strong it may be, is a finite quantity, whereas the power that we can bring to bear against it is the infinite power of the human will, which can make renewed efforts day after day, year after year, even life after life if necessary.

Another great difficulty in our way is the lack of clearness in our thought. People in the West are little used to clear thought with regard to religious matters. Everything is vague and nebulous. For occult development vagueness and nebulosity will not do. Our conceptions must be clear-cut and our thought-images definite. Other necessary characteristics are calmness and cheerfulness; these are rare in modern life, but are absolute essentials for the work which we are here undertaking.

The process of building a character is as scientific as that of developing one's muscles. Many a man, finding himself with certain muscles flabby and powerless takes that as his natural condition, and regards their weakness as a kind of destiny imposed upon him; but anyone who understands a little of the human body is aware that by continued exercise those muscles can be brought into a state of health and the whole body eventually put in order.

In exactly the same way, many a man finds himself possessed of a bad temper or a tendency to avarice or suspicion or self-indulgence, and when in consequence of any of these vices he commits some great mistake or does some great harm he offers it as *an* excuse that he is a hasty-tempered man, or that he possesses this or that quality by nature—implying that therefore he cannot help it.

In this case just as in the other the remedy is in his own hands. Regular exercise of the right kind will develop a certain muscle, and regular mental exercise of the right kind will develop a missing quality in a man's character. The ordinary man does not realize that he can do this, and even if he sees that he can do it, he does not see why he should, for it means much effort and much self-repression. He knows of no adequate motive for undertaking a task so laborious and painful. The motive is supplied by the knowledge of the truth.

One who gains an intelligent comprehension of the direction of evolution feels it not only his interest but his privilege and his delight to co-operate with it. One who wills the end wills also the means; in order to be able to do good work for the world he must develop within himself the necessary strength and the necessary qualities. Therefore he who wishes to reform the world must first of all reform himself. He must learn to give up altogether the attitude of insisting upon rights, and must devote himself utterly to the most earnest performance of his duties. He must learn to regard every connection with his fellow-man as an opportunity to help that fellow-man, or in some way to do him good.

One who studies these subjects intelligently cannot but realize the tremendous power of thought, and the necessity for its efficient control. All action springs from thought, for even when it is done (as we say) without thought, it is the instinctive expression of the thoughts, desires and feelings which the man has allowed to grow luxuriantly within himself in earlier days.

The wise man, therefore, will watch his thought with the greatest of care, for in it he possesses a powerful instrument, for the right use of which he is responsible. It is his duty to govern his thought, lest it should be allowed to run riot and to do evil to himself, and to others; it is his duty also to develop his thought-power, because by means of it a vast amount of actual and active good can be done.

Thus controlling his thought and his action, thus eliminating from himself all evil and unfolding in himself all good qualities, the man presently raises himself far above the level of his fellows, and stands out conspicuously among them as one who is working on the side of good as against evil, of evolution as against stagnation.

The Members of the great Hierarchy, in whose hands is the evolution of the world, are watching always for such men in order that They may train them to help in the great work. Such a man inevitably attracts Their attention, and They begin to use him as an instrument in Their work. If he proves himself a good and efficient instrument, presently They will offer him definite training as an apprentice, that by helping Them in the world-business which They have to do he may some day become even as They are, and join the mighty Brotherhood to which They belong.

But for an honour so great as this mere ordinary goodness will not suffice. True, a man must be good first of all, or it would be hopeless to think of using him, but in addition to being good he must be wise and strong. What is needed is not merely a good man, but a great spiritual power. Not only must the candidate have cast aside all ordinary weaknesses but he must have acquired strong positive qualities before he can offer himself to Them with any hope that he will be accepted.

He must live no longer as a blundering and selfish personality, but as an intelligent ego who comprehends the part which he has to play in the great scheme of the universe. He must have forgotten him-self utterly; he must have resigned all thought of worldly profit or pleasure or advancement; he must be willing to sacrifice everything, and himself first of all, for the sake of the work that has to be done. He may be *in* the world, but he must not be *of the* world. He must be careless utterly of its opinion.

For the sake of helping man he must make himself something more than man. Radiant, rejoicing, strong, he must live but for the sake of others and to be an expression of the love of God in the world. A high ideal, yet not too high; possible, because there are men who have achieved it. When a man has succeeded in unfolding his latent possibilities so far that he attracts the attention of the Masters of the Wisdom, one of Them will probably receive him as an apprentice upon probation.

The period of probation is usually seven years, but may be either shortened or lengthened at the discretion of the Master. At the end of that time, if his work has been satisfactory, he becomes what it commonly called the accepted pupil. This brings him into close relations with his Master, so that the vibrations of the latter constantly play upon him, and he gradually learns to look at everything as the Master looks at it. After yet another interval, if he proves himself entirely worthy, he may be drawn into a still closer relationship, when he is called the son of the Master.

These three stages mark his relationship to his own Master only, not to the Brotherhood as a whole. The Brotherhood admits a man to its ranks only when he has fitted himself to pass the first of the great Initiations.

This entry into the Brotherhood of Those who rule the world may be thought of as the third of the great critical points in man's evolution. The first of these is when he becomes man—when he individualizes out of the animal kingdom and obtains a causal body. The second is what is called by the Christian "conversion", by the Hindu "the acquirement of discrimination", and by the Buddhist "the opening of the doors of the mind". That is the point at which he realizes the great facts of life, and turns away from the pursuit of selfish ends in order to move intentionally along with the great current of evolution in obedience to the divine Will.

The third point is the most important of all, for the Initiation which admits him to the ranks of the Brotherhood also insures him against the possibility of failure to fulfil the divine purpose in the time appointed for it. Hence those who have reached this point are called in the Christian system the "elect", the "saved" or the "safe", and in the Buddhist scheme "those who have entered on the stream". For those who have reached this point have made themselves absolutely certain of reaching a further point also—that of Adeptship, at which they pass into a type of evolution which is definitely Superhuman.

The man who has become an Adept has fulfilled the divine Will so far as this chain of worlds is concerned. He has reached, even already at the midmost point of the aeon of evolution, the stage prescribed for man's attainment at the end of it. Therefore he is at liberty to spend the remainder of that time either in helping his fellow-men or in even more splendid work in connection with other and higher evolutions. He who has not yet been initiated is still in danger of being left behind by our present wave of evolution, and dropping into the next one—the "sonian condemnation" of which the Christ spoke, which has been mistranslated "eternal damnation". It is from this fate of possible sonian failure—that is, failure for this age, or dispensation, or life-wave—that the man who attains Initiation is "safe". He has "entered upon the stream" which now *must* bear him on to Adeptship in this present age, though it is still possible for him by his actions to hasten or delay his progress along the Path which he is treading.

That first Initiation corresponds to the matriculation which admits a man to a University, and the attainment of Adeptship to the taking of a degree at the end of a course. Continuing the simile, there are three intermediate examinations, which are usually spoken of as the second, third, and fourth Initiations, Adeptship being the fifth. A general idea of the line of this higher evolution may be obtained by studying the list of what are called in Buddhist books "the fetters" which must be cast off—the qualities of which a man must rid him-self as he treads this Path.

These are: the delusion of separateness; doubt or uncertainty; superstition; attachment to enjoyment; the possibility of hatred; desire for life, either in this or the higher worlds; pride; agitation or irritability; and ignorance. The man who reaches the Adept level has exhausted all the possibilities of moral development, and so the future evolution which still lies before him can only mean still wider knowledge and still more wonderful spiritual powers.

The Four Stages of Life

Hinduism recognizes four main stages of life. Like the goals of life, these can be divided into three plus one, with the three deriving from the "life is good" strand of Hinduism, and the one deriving from the "life is bad" strand. The first three are the student, the householder, and the retired person, while the fourth is the ascetic (also known as a sannyasin or a sadhu).

Progressing Through Life: The Three Stages

The three stages of life that come from the life-affirming, Vedic side of Hinduism, were initially designed with the caste system in mind (of course). In particular, they were set up to apply to members of the three Twice-Born varnas: the Brahmin, the Kshatriya and the Vaishya. Other castes and jatis have adopted them in different ways, transforming them to meet their needs. The first stage is that of the student, during which a boy traditionally is expected to go to live and study with a teacher (a guru) for several years. Today only a few Brahmin families follow this tradition to the full extent. A boy enters into student-hood at adolescence (ages 8-12), and spends most of his maturing years studying. For Brahmins, this would mean studying and memorizing large portions of the Vedas and accompanying texts, along with training in the various rituals. Members of all castes learn how to set up and maintain their own household worship, centred on the holy fire of Agni.

For the Twice-Born castes, the ritual (samskara) of becoming a student contains great significance, for it is the means by which a person becomes reborn. This ceremony—often called the thread ceremony because of the red thread which the initiate wears over his left shoulder—symbolizes the entrance of the boy into Hinduism. Originally, it was at this point that the initiate was first permitted to hear the words of the sacred Vedas and learned his first mantra. Once initiated, the boy became, like other Twice-Born males, responsible for maintaining the balance of the cosmos.

After student-hood, the next stage of life is that of householder, usually entered into through an elaborate, many-day marriage ceremony. It is during this stage that a man has children (with his wife), forms a family, establishes himself in a career or job, and strives to be an active member of his community. He will establish his own household, with its own worship. Indeed, with his wife, the householder is now responsible for ensuring that the rituals of domestic life are carried out at their proper times and in the proper manner. This stage is important because it carries the responsibilities of looking after and supporting people at all other stages, both male and female.

The third stage of life is that of retirement. When a man reaches old age and his son has a family and is ready to take over the leadership of the household, he and his wife will retire. On the one hand, their household responsibilities—both religious and secular—diminish significantly. On the other hand, they become free to contemplate the meaning of their coming death and rebirth. They may choose to withdraw into a secluded area—perhaps become a "hermit"—or they may involve themselves in more active worship (bakti) of Hinduism's pantheon of gods and goddesses.

Each of these three stages is preceded by a samskara, a ritual that brings a person from the previous stage of life into the new one. While these are the most important stages of life, brought on by the most elaborate samskaras, there are many other samskaras performed during one's life. Traditionally, a person may undergo anywhere from 10 to 18, even up to 40, samskaras during their lifetime. The majority of these will be performed before a baby is even six months old, with many of them done before birth. These are believed to help a person leave their previous life behind and to enter successfully into their new one. Each samskara advances a person further along the path of life, initiating them into a new aspect or stage.

It is clear that the three main stages of life are designed for males and do not include women. Traditional Hinduism, like many religions, places women in a dependent role. In the traditional view, women always need the protection of a responsible male, whether father, husband or adult son. This does not mean that women have no religious life. On the contrary, women are actively involved in worship, both in support of their family and on their own.

On the one hand, a married woman is responsible for carrying out many of the domestic rites along with her husband. Many rituals cannot be performed with her involvement or in some instances leadership. On the other hand, women are often active practioners of forms of bakhti yoga, that is, the worship of the gods and goddesses. In the modern period, this subordination has begun to change and women have gained more active roles in public life. Indira Ghandi, for instance, was a Prime Minister of India for many years (women have yet to gain a corresponding position in the United States).

Rejecting Life: The Fourth Stage

The fourth stage of life breaks the progression of the other three; it is that of the ascetic, who in Hinduism are called the sadhu or the sannyasin. This is a rejection of life and all that it means in exchange for a search to attain moksha, that is, release from the cycle of samsara. A person may enter into this stage of life at any time.

The rejection of life, especially as defined by the life-affirming strand of Hinduism, is complete. It requires rejection of the household duties and responsibilities of all stages of life. It also requires the rejection of the religious beliefs. Indeed, the ceremony making one a sannyasin includes the burning of copies of the Vedas, a symbolic rejection even of one's role in maintaining the cosmos, and of one's red thread, the symbol of their status as Twice-Born.

It is such a powerful rejection that a person even loses their caste affiliation; even a shudra can become a sannyasin and lose their low-caste identity.

The sannyasins become wandering hermits, living life without any shelter or possessions. They eat when they can acquire food, but never enter into any work to acquire it; it must be given or found. They become holy men, seeking spiritual enlightenment and power, striving to achieve the true wisdom of the cosmos. Some may become kind and give blessings to those around them, while others may become wrathful and powerful and wield magic against those who cross them.

The Four Stages of Life

I. The Four Stages of Life—as has been noted before, if there is one abiding perspective on Hinduism, it is that people are different. How should you live? If you are a male in an upper three *varna*, then it depends upon what stage of life you are living. (Not all persons go through all stages).

   A. The Student Stage—(twelve year rite of initiation 8,12 to 18,24 years old for male, high caste Hindus).

      1. Student lives in the home of his teacher and study the Vedas.
      2. Student serves the teacher, shows respect, and learns the texts. The rite is a kind of apprenticeship where habits, skill, and practical knowledge are emphasized.

   B. The Householder Stage—(begins with marriage as the completion of the formal studies).

      1. The rules for marriage are set forth in the *Laws of Manu*. Attention is turned toward the world: family, career, and community.
      2. Note the relation to the game of life, the path of desire, where success is a means to self-esteem.
      3. If physical activity is the only importance in life, all experience after youth is wasted, and

what would lie ahead would be unfortunate. However, obviously, there is more to life than *dharma*.

C. The Forest Retirement Stage—(begins after the arrival of the first grandchild; the grandfather can withdraw from social obligations).

1. Compare the different import to the question, "What do you want to be?" when the question is addressed to a 16 year-old and to a 50 year-old person.
2. This stage involves coming to terms with who we are. There is time to read, think, and consider the significance of life without the interruption of duty.

D. The Forest Dweller or Ascetic Stage—(begins by leaving home and carrying out a spiritual existence in the country).

1. The man and his wife together (if she wants to go) move to the forest to begin in earnest the path of self-discovery.
2. Most men defer the Forest Dweller Stage to another future life.
3. The forest dweller works out a philosophy of *sannyasin*—one who neither hates nor loves anything. A *sannyasin* is completely independent and is beyond dharma (the structure of moral and social obligations) and so in a sense is "beyond good and evil."

a. There are no social pretensions—things simply are what they are (*cf.*, Vasudeva in Hesse's *Siddhartha*).

b. Once detachment, mental and economic independence, is achieved, the *sannyasin* can return to the town or city.

c. This stage of life is a necessary condition for the attainment of salvation; once achieved that soul will never individually return to this world.

Religious Life of a Hindu

For a Hindu everything in life is religion: Family, work, play, walk you breed religion in India! Daily worship in Hinduism usually takes place in three different places: in the home, in a temple, and/or at a street-side or road-side shrine.

At Home: Most religious life is at Home: There they have a room for "puja" (prayer), with an image or picture of the favourite god, and a "mandala", a symbolic representation of the Universe. They use the robe and the distinctive in the forehead of the caste, or the god they adore: Vishnu adorers use 3 perpendicular lines in the forehead, Shiva worshippers, use 3 parallel lines...

They start the morning with the "mantra" "OM", pronounced "AUM", (That Who Art); then they invoke the favourite god, repeat a mantra from the Vedas... all of this sitting down, nude to the waist, with the face towards the sun. Then they drink a little water, touch 6 parts of their bodies, indicating the god is inside them; offer water to the images, repeat the mantra from the Vedas. At night is similar. And at noon may include a consultation with a teacher (guru).

At dawn, the householder and his wife rise, purify themselves with a bath—usually in a temple pool or a river if one is available—and then make an offering to the fire-god Agni in their household fire. The man may then turn towards the rising sun and say a mantra to the sun-god Savatar, asking for blessing and understanding. A similar sequence of activities will take place in the evening.

Second, most Hindu households have a small shrine to the gods important to that house. It may have a small statue of Krishna or a picture of Shiva or Durga. If the householder has a guru, a photo of the guru will appear, to remind the worshipper of the guru's teachings. This shrine will be the focus of household puja, *i.e.*, worship. Offerings of food or drink may be laid before the statues, mantras and prayers may be said, and so on.

At the Temple: A nearby temple to a god or goddess is usually the focus of regular puja (*i.e.*, worship). While a local temple may do for everyday worship, a grander cathedral-like, temple may be visited on special occasions. Offerings of meals, money, flowers, etc. may be brought by the devotee. Once the god has taken his part of the sacrifice, the devotee may share in some of the now-blessed food (called prasad). The worshipper may also say mantras, or listen to the priests chant, sing, or read from the sacred texts.

Small shrines to Hindu gods and goddesses, both major and minor, stand on road sides in the country and on the streets in cities. They may be permanently fixed and unattended, or on a cart and moved around by an attendant. During the day, as people pass by, they may stop, offer a short prayer or mantra, perhaps leave a small offering in gratitude for some blessing.

The "priests" take care of temples and the rituals in the temple or at home, with very complicated and long beautiful rituals. The "Swamis", usually a learned celibate monk; the "Shadus", are less educated holy men who seek spiritual merit meandering restlessly across India, and often nude with dung of a cow in the head. The "guru" is a teacher who has mastered the path of yoga, and is like a god, expects honours and obedience as a god.

And then, after the gods and the priestly system, come the adoration of sacred rats, cobras... and very specially the "sacred cows", all 159 million of them; they freely roam the urban centres, like Bombay and Calcutta, and they have "rest homes" for the old or sick cows.

Women in Hindu Scriptures

Hinduism is a patriarchal religion, but it is ridiculous to say women there have no rights or were constantly oppressed. Unfortunately most foreigners seem to associate Hindu women with Sati and non-remarriage of widows and many Hindus remain unaware of what their own scriptures said. Yet, it is very easy to trace how restriction on women tightened until they became oppressive.

So I decided to put down aspects of women in Hindu religious texts that I find personally fascinating. If anyone has hyper-sensitive moral antennae or is over-civilized please go away.

- Polyandry in early Hindu society.
- Eight types of Hindu marriages.
- Twelve types of son in Hinduism.
- Child-Marriage.

Polyandry in Early Hindu Society

In Vedic society marriage was naturally a matter of supreme importance, but there does not seem to be any fixed rules regarding it. Though monogamy no doubt existed, the society was polygamous. The Vedas contain prayers to defeat rival co-wives and place the husband under the speaker's control. But evidence indicates that polyandry was also common. Many scholars in the past, both Indians and Europeans, have been reluctant to accept this, and consequently they sometimes translated the plural form as the singular. However, more proper translations show that there is enough indication of polyandry.

The first evidence lies in the metaphors used. Three images stand out. Surya, daughter of sungod, is constantly described as the bride of twin Asvins. Rigveda, I, 116, 17 declares that they competed in a divine chariot race for her hand and both won her. Elsewhere (I, 119, 5) says, "The virgin Surya after being won, out of friendship declared "Thou (plural form is used in the original) are my *husbands*" and elected you as her lords". Another verse says that she had another husband, Soma, and Aswins were also her husbands at the same ceremony (X, 85, 9-15). Even if they are only metaphors of natural forces—as they often are—the poets wouldn't have introduced such a reference in sacred writings if it was not practiced among mankind as well.

The Maruts or stormgods are said to have one wife Rodasi, their 'common wife'. Rodasi tends to the Maruts in order to be united sexually with them. She mounts their chariot and hastens with them to come to the sacrificial altar. It is clearly stated that she is their joint wife. (I, 167, 4-8).

Though it is said that Maruts are storm clouds and Rodasi lightning, there is no reason for the poet to call her their common bride and them her husbands, when sister or daughter would have been an equally descriptive term. Instead the term is 'sadharani' or common wife. Similarly, VII, 33, 11 hail the sage Vashistha as the son of both Mitra and Varuna from Urvasi. He thus has two fathers but one mother. This of course, can be explained as another metaphor, but it shows that this was not thought to be something shameful.

The above references, could be explained as metaphors. However in many of the verses, the reference is explicit. The marriage hymn, Rigveda X, 85,37-38 says: "O Pusha [sungod]! The woman in whose womb men plant their seeds, send her as a propitous one. She, ruled by desire yields her body, *we* embrace her with desire ... you [fire] give the bride with children to *husbands*." The verses are explained as declaring that husbands referred to here are the gods who are guardians of the girl (like guardian angels), but surely the plural terms used are significant. When we look at Atharva Veda nuptial hymns, we see the same thing.

In Atharvaveda, XIV, 1:44, the bride is being blessed thus, "Be thou supreme among *fathers-in-law*, supreme also among *brothers-in-laws*; be thou supreme over *sister-in-law*, supreme also over *mother-in-law*". Again, XIV, 1: 46 declares, "Joy to the *husbands* embracing the *wife*".

There are several other mentions of Fathers in this context, but they can have other meanings. However these two verses are certainly proof of polyandry. Both Taiterya Samhita and Aitarya Brahmana (later portions of the Vedas) forbid polyandry; they declare that a man can have more than one wife but a woman cannot have more than one husband. Obviously such a regulation would not have to be passed if the practice was unknown.

Polyandry can also explain why in later ages, the younger brother-in-law was considered to be the proper husband for a widow. The very term for a younger brother-in-law is 'devar'— meaning second husband. This could either refer to the fact that through marriage of the elder brother all brothers became married at once, or that he automatically inherited the post of husband as his closest kin, if the elder died.

If we accept that polyandry was practiced in early Indian society then, it explains the only concrete example of it—Draupadi's marriage in the epic Mahabharata. The incidents of the epic takes place considerably later than the Vedic period, when civilization had settled down and wealthy and powerful kingdoms have appeared. Naturally restrictions on women have also tightened.

Kunti the mother of Pandavas, had abandoned her first-born because he was illegitimate. Yet, Yuddhisthir, her eldest Pandava son, decides that all five brothers shall marry Draupadi. At first Draupadi's father refuses saying that while a man can have many wives, it has never been known for a woman to have many husbands.

At this Yudhishthira gives examples from the past: Jatila had married seven sages and Barkhi had been the wife of ten brothers. The marriage duly takes place and Draupadi becomes the wife of all. There was no reason for Yudhishthira to have made such a suggestion unless it was already known. Similarly, society at large would never have accepted it—though the enemies of the Pandavas mockingly call her no better than a courtesan—unless it was an established custom.

We are told that all people of the city including the Brahmins happily attended the wedding and Vedic rites were performed. Interestingly, the wedding rite takes place with only Yudhishthira, the eldest brother. Nevertheless, she became the legal wife of the younger brothers as well. Thus her brother-in-laws truly became her 'devars' or second husbands. Evidently during this epic period polyandry was disappearing due to disapproval of patriarchy, but it was still known and grudgingly accepted.

Eight Types of Hindu Marriages

Concurrent with the epic period, we enter the period of Dharmashastras or sacred lawbooks. These Shastras are not as holy as Vedic literature and here dharma means not 'religion' so much, but the dos and don'ts of society. It is not that they give new laws, they simply codify what was in usage, try to fill up some gaps and thereby grant the laws scriptural authority.

They are considered to be location and time-specific and so can be ignored, whereas Vedic literature and the two epics are eternal. Many of them—particularly the most famous one, called Manusamhita,—had been edited over time, so that it is hard to fix their chronology or get an entirely accurate picture. Many Hindus have only heard about them but regrettably had not read them. One of their most fascinating features is their classifications of marriages and children. More or less, they all agree on the following:

1. The gift of a daughter, after decking her with as much costly garments and jewels as possible, to a learned man of good conduct, whom the father himself invites, is called the Brahma rite.
2. The gift of a daughter who has been decked with ornaments, to a priest who duly officiates at a sacrifice, during the course of its performance, is called the Daiva rite. [here the hope is that by such a marriage the priest's prayers would be doubly powerful].
3. When the father gives away his daughter according to the rule, after receiving from the bridegroom according to sacred laws, a cow and a bull or two pairs of cows, that is named the Arsha rite.
4. When a contract is signed with the husband, and it has been said [by the bride's father] that "You two shall now practice your [householder's] duties together" and the daughter is given after the husband had been duly honoured, then the marriage is known as Pragyapata marriage. [here possibly, the husband himself sought the girl's hand].
5. When the bridegroom receives a maiden, after having given as much wealth as he can afford to the kinsmen of the bride, and to the bride herself, according to his own will, that is called the Asura rite. [here instead of the husband receiving the dowry, it is the bride and her family who receive it].
6. The voluntary union of a woman and her lover is the Gandharva rite, which springs from desire and has sexual intercourse for its purpose. [Here, no rituals or

witnesses are necessary; merely the couple exchanging vows is sufficient].

7. The forcible abduction of a maiden from her home, while she cries out and weeps, after her kinsmen have been slain or wounded and their houses broken open, or during war, and subsequent marriage is called the Rakshasa rite.
8. When a man by stealth seduces a girl who is sleeping, intoxicated, or disordered in intellect, and then marries her, that is the eighth, the most base and sinful rite of the Pisakas.

From the above laws we see that various kinds of marriages were prevalent in those days. The first four were called 'dharmavivaha' [holy marriages] since they had been contracted according to proper sacred rituals with the aim of procreation. Marriage in these cases is a social and religious contract. The Asura rite was condemned as nothing other than sale of daughters and it was strictly forbidden to anyone desirous of escaping hell.

The Gandharva or love marriage is called ordinary or common marriage as contrasted with 'dharmavivaha'. The most famous example of this type of marriage is the marriage between King Dushmanta and the forest-dwelling Shakuntala, where it is sealed by their wishes alone. Here marriage is an intensely personal matter and has nothing to do with family obligations. Possibly that is why the writers of the shastras did not wholly approve of it and declared that in case of such a marriage, the bride's parents are not obliged to give the bride any gifts. The 7 marriage, Rakshasha, is condemned as barbaric; however there seems some confusion. Some shastras declare that it is the *only* lawful marriage for the warrior-caste while others argued that it is also 'dharmavivaha' for the warrior-caste.

The implication is obvious. In the distant past, warriors frequently raided other tribes or cities and carried away women as brides—the verses sanctifying such marriages therefore are earlier than other verses which condemn this. Pikasas marriage is condemned by all. In later ages, the shastras, declared that only the first six are valid. Their preference is for the first four alone. But they also realized they cannot control human nature: wealth will lure fathers to give away their daughters and many daughters would prefer a wealthy suitor to a learned one; men and women will fall in love and marry in defiance of authority and duty. There is therefore nothing to do but recognize them and the children born of these unions. Though the last two types of marriages were condemned in scriptures, social reality was otherwise.

Kautilya's Arthashastra, (a secular lawbook written for the administration of the Mauryan empire and later a manual for government for other Hindu states), recognizes even the last two as legally valid. Marriage was held to be a sacrament. Some declared that a girl who had been forced was like a virgin and could be given in marriage to another; but the qualifying condition was she must be unmarried. A marriage performed with proper ceremonies, even if the bride was under duress, cannot be dissolved, and additionally provided legitimacy for children born from such unions. However, Arthashastra does allow divorce in the case of last four marriages (but not for first four) if both parties agree to it.

Twelve Types of Sons

Various categories of children are legitimized and entitled to inheritance of the 'father's' estate, in the shastras and the epics, which I can confidently claim had not been sanctioned by any other culture.

1. Him whom a man begets on his own wedded wife, is the son of the body (Ourasa), the first in rank.
2. He who was begotten according to the law of Niyoga (when the husband is dead, impotent, diseased, or missing, the relatives of the wife's or husband's family, or sometimes the husband himself, would designate someone to make the wife pregnant) is called a son begotten on a wife (Kshetraga). [this was a very widespread practice and many husbands who are otherwise able, willingly sent their wives to men they considered to be superior. Several heroes and kings

were born in such a way who inherited kingdoms from their mother's husband. The Jews and Spartans too had this kind of system. This however becomes oppressive when the woman is unwilling, as some stories testify where the women send maidservants in their stead]

3. That boy equal by caste whom his mother or his father affectionately give, or gave in times of distress, to a man as his son, is the adopted son (Dattak).
4. When a boy who is capable of judging between right and wrong, and endowed with filial virtues, belonging to the same caste is accepted as a son, he is called artificial (krittim) son. [here the boy is obviously mature enough and his parents do not seem to have any role to play].
5. He whom a man receives as his son, [of same caste or not] after he has been deserted by his parents or by either of them, is called a son cast off (Apaviddha).
6. If a man buys a boy, whether equal or unequal in caste, from his father and mother for the sake of having a son, that son is called a son bought (Kritaka).
7. He who, having lost his parents or being abandoned by them without (just) cause, gives himself to a man, as his son, is called a son self-given (Svayamdatta)
8. If a woman abandoned by her husband, or a widow, of her own accord contracts a second marriage and bears a son, he is called the son of a re-married woman (Paunarbhava).
9. If a child be born in a man's house and his father be not known, [*i.e.*, he is born not from niyoga but from his mother having an affair only], he is a son born secretly in the house (Gudhotpanna or grihaja), and shall be the legitimate son of the husband of the mother.
10. A son whom a damsel bears in the house of her father, is the son of an unmarried woman (Kanina) and such off-springs of an unmarried girl belong to him who weds her afterwards.

11. If one marries, either knowingly or unknowingly, a pregnant bride, the child in her womb belongs to him who weds her, and is called a son received with the bride (Sahodha).
12. If there is no son, a daughter can be appointed as a son (Putrika) or the daughter's son shall be designated as a son (Putrika-Putra). [such designations can be done through formal rituals, or some say, will be valid simply by wishing it]

We thus see the ancient scripture-writers were very generous because they accepted human nature; practically every child born in whatever way was considered legitimate and entitled to a share in the husband's inheritance. Naturally the biological son and the daughter/daughter's son, were considered to have the greatest claim, with the Khetraga usually coming next. Depending it seems on the age of the shastra, the ranks of the sons vary, with some arguing that a wife's sons are higher than an adopted son or that only some kind of sons can actually inherit the husband's property, the rest being only kinsmen.

Arthashastra too records these various types of children and their claims on property. However, as time passed and women's status degenerated with restrictions imposed on their sexuality, society became less accepting of such affairs. Moreover, these various kinds of sons led to intermixing of castes, with the children born in different castes inheriting the caste of the legal father. Finally, all kinds of sons, except the son of the body and adopted sons and grandsons by a brotherless daughter were forbidden by custom, and this injunction had been followed by Hindu society since.

The above marriages and sons probably explain the development of child-marriages. Vedic society knows nothing of child-marriages, nor is it mentioned anywhere in the epics. On the contrary all marriages are between men and women who had attained puberty. But as we can see from evidence of love-marriages and 'sons of unmarried women', that adolescent girls were sexually active in their natal homes. Fear of scandal was a strong impetus in marrying off girls as quickly as possible, so that the shastras insisted that it is best if a girl is married even before menstruation.

The choice of husbands was possibly another important factor. The same shastras agree that "three years after menstruation, a maiden gains overlordship over herself. She is then free to select her own husband and her parents can have no say, since they had failed to find a suitable match for her". This meant she often married someone whom her parents would disapprove of. There is a special term for people who married out of their varna: Anuloma marriage is when a man of higher caste marries a woman of lower caste, and Pratiloma marriage is one where a woman of higher caste marries a man of lower caste. Among conservative families, especially among Brahmins, Pratiloma would naturally be regarded with abhorrence.

Though the practice seems to have been extensive, scriptures grew more and more harsh towards such unions, declaring that untouchables are born this way, until it stopped. Yet another kind of marriage is 'Swayamvara' (selecting one's husband herself), practiced mostly among the warrior-castes and royal families. A group of suitable candidates would turn up and the bride would choose among them. Probably she had the right to refuse to marry any as Draupadi refuses to marry Karna, but it is not certain.

In historical times, Princess Sanjukta chose in this way Prithviraj, whom her father considered his mortal enemy, though both were kings. Incidents like these obviously made the parents marry off their daughters still as a child. In the middle ages, as society became more rigid, child-marriages flourished. Blind reverence for scriptures and customs prevented parents from questioning it. Waves of Islamic invasions probably contributed to this impulse as well; if the heathens do carry off your daughter, it is comforting to know that the 'disgrace' is now the headache of the husband's family.

Hindu Sacred Thread

"In the eighth year from the conception of a Brahman, in the eleventh from that of a Kshatriya, and in the twelfth from that of a Vaisya, let the father invest the child with the mark of his class." (Mann, ii. 36.)

One of the many peculiarities that strike a stranger in India is that many Hindus have a cord or skein of thread over the left shoulder, hanging down under the right arm. It is worn as a sash would be. Probably few, except the Hindus themselves, could tell why this cord is worn; why certain persons have it whilst others have it not; or even how or of what it is made.

This article of dress or adornment forms, however, a very important factor in the Hindu cult. The yajno pavitam, as it is 'called, or the sacred thread of the Hindu, is the outward and visible mark that the wearer is a Dvija, or twice-born. It is a very much prized and a very sacred badge and commands respect and even adoration.

If we enquire who are privileged to assume this distinction, we find the matter very clearly defined by the ancient Hindu law-giver. In the quotation at the head of this chapter, it is clear that the Brahmans, Kshatriyas, and Vaisyas must be thus invested. In another place it is distinctly stated that none but the three twice-born classes are entitled to the distinction.

"The three twice-born classes are the sacerdotal, the military, and the commercial; but the fourth, or servile, is once-born, that is, has no second birth from the gayatri, and wears no thread." (Manu, x. 4.)

This is the law; but others besides these three privileged classes assume the distinction. It is not safe, therefore, to conclude that every wearer of the sacred thread must necessarily be a Brahmin or one of the other two highest castes. The goldsmiths, the weavers, certain classes of fishermen and others wear it. The explanation is that the goldsmith caste, many of whom are carpenters and workers in brass and copper, are themselves a class of Brahmins; at least they assume the distinction. They have their own prescribed share in the Vedas and their own ritual. They have an upanayanam or second-birth ceremony, and are considered dvijas, or twice-born; hence this privilege in the matter of the thread, as well as in many others that are peculiar to Brahmins.

The goldsmith caste are said to be the descendants of Brahmin women and Kshatriya men; and this fact, together with the rights above mentioned, appears to be acknowledged by the Brahmins themselves; yet, they do not appear to command much respect as a caste. This may be on account of their mixed origin. Until recent years, for instance, they were not allowed to celebrate their marriages with public processions, to use a palanquin, or to ride a horse. It is said that about thirty years ago there was much disturbance in Masulipatam when, in India through the freedom resulting from the British rule this caste first began to have marriage processions and, in other ways, to assert themselves.

Now it is quite an acknowledged thing, for it has become what is known as ma'mul (custom) and so no one interferes. They are not, as a rule, even now, allowed to enter temples. When they are permitted to do so, it is only to that part in which Śudras are allowed. A case recently came before the law courts in Masulipatam arising out of the attempt of a goldsmith to enter a Shiva temple for worship. I believe it was eventually decided that a member of this caste could not enter a temple, except by the permission of the Brahmin priest in charge. The question of the social and religious status of the goldsmith caste is a most vexed one which gives rise to much controversy. It shows that not all wearers of the thread or cord are considered of equal rank.

I now make a slight digression to say a little as to the denial of the liberty of an individual to dress or to go as he pleases. There are very binding rules and regulations on these points. They are the outcome of caste customs, which, whatever may be the real rights of the matter from a legal point of view, are very real and strict in actual life. Theoretically, for example, any British subject has the right to use the public road in the way and manner of others, whatever may be his degree; but, practically, this is not so. A low caste man, in going through a respectable public street, inhabited by high-caste people, must take off his shoes and turban and shut up his umbrella, and, if he should be riding, he must descend from his horse and humbly walk through on foot.

Even if a Sudra should be riding and happen to pass a superior person, as a Pandit, or high official, he must descend and walk past on foot until he is well past the dignitary, when he may remount and go on his way. A case quite recently came before the law court in a district where a native Pastor of one of the Christian communities was severely beaten, because he dared to ride through a respectable public thoroughfare.

As a man of low caste origin, he should have humbly descended and have gone through on foot. To return, however, to the subject more immediately in hand, others such as weavers and fishermen appear unlawfully to assume the privilege of wearing the sacred thread; but, although their custom is not interfered with, no value is set upon it by orthodox Hindus. They cannot, for example, read the Vedas or even hear them read. Authority to do this, in the case of lawful thread-wearers, is conveyed by the ceremony of upanayanam, or second spiritual birth, of which the thread is the outward symbol.

In these modern days, some other classes of Sudras have also adopted the yajnopavitam merely to add to their own importance; but, in all such cases, it is of no true religious value. I have heard of a case in the Orissa country where a certain Raja of the Sudra caste made himself important by assuming authority to invest people of his own caste with the thread. Some of, them, to please him, appear to have submitted to the investiture, and adopted the thread—thus adding to the number of the unlawful wearers of this coveted mark of distinction. It is said that one unlucky weight who, on a visit to that country, was presented with this badge of honour, was, on his return home, deprived of the same and well beaten for his presumption by his indignant neighbours.

Having thus seen who are entitled to wear the sacred thread, I now pass on to mention some particulars of the thread itself and also of the mode of investiture. Originally there appears to have been some difference in the kind of thread worn, according to the class of the wearer. Thus:

"The sacrificial thread of a Brahmin must be made of cotton, so as to be put on over his head in three strings;

> *that of a Kshatriya of sana thread only; that of a Vaisya of woollen thread." (Maim, ii. 44.)*

This is the law and probably in ancient times the material of which the thread was made did thus differ, according to the caste of the wearer; but in the present day no such difference is seen. The cord is universally made of cotton. A peculiar kind of very fine cotton is what ought to be employed, but ordinarily the common cotton is used. The threads are supposed to be prepared by Brahmins. Perhaps other than the Brahmins and Vaishyas are not so particular as to the manufacture, but these two castes are very careful in this respect. The threads can generally be obtained in any ordinary bazaar, but the very orthodox, in order to ensure their purity, will frequently procure their supply from the house of the Brahmins who may happen to be engaged in the manufacture.

The yajnopavitam consists of several skeins of cotton thread. Each thread consists of three strands, each skein has three threads, and a married man's cord must consist of not less than three skeins. The number three enters very largely into the structure of the cord itself, and the ceremony of investiture.

This is said to represent the three gods, Brahma, Vishnu, and Shiva; the three sacrificial fires; three divisions of time — morn, noon and night; and the three worlds—heaven, earth and hell. Each skein is tied with a peculiar knot called Brahma's knot. It is made by making three turns with the threads and by so tying the knot that the ends do not appear on the outside. In making each knot the following incantation is repeated by the maker:

> *"Pronouncing the word Om, the Brahma Sutram must be tied, and afterwards worn. (The wearer) will receive purity in all his rites, It being the personification of all the gods."*

A youth, if a bachelor, when first invested with the cord, receives only a single skein, and he cannot wear more than a single skein until he is married, when he must wear, at least, three skeins. The Brahmin youth must be invested with his cord when he is about seven or eight years of age. He cannot be married until thus invested, but he may, and in fact often does, marry a day or two after the ceremony. Amongst some of the Banians or Vaishyas, it is customary to defer the upanayanam until immediately before marriage.

The ceremony of investiture is as follows. On the appointed day a fire is lighted, round which the relatives and friends of the novice are seated. This fire is a very important feature of the upanayanam. The whole ceremony is called the agni karyam or fire worship. It is kept alight during the whole four days during which the ceremony lasts, and it is the proper thing to feed it, as far as possible, with the twigs of certain kinds of trees,—principally those of the Indian fig tribe. At the repeating of the various mantrams which form part of the ritual, ghee is poured on to the fire as an offering. The father of the youth to be invested takes a thread of nine strands and puts it upon his son.

This is not the true yajnopavitam, nor has it the Brahma knot, neither are mantrams said over it. After some time, during which various rites are performed, and the ears of the boy are bored for earrings and then adorned with thin rings of gold, the true cord is produced—a single skein of three threads. To this is attached a bit of the skin of a male deer, or, if procurable, a long strip of this skin is worn as a sash together with the cord. Deer skin is considered to be very pure, and also to be capable of imparting purity. For this reason untanned deer skin is much employed for covering the boxes and other receptacles, in which the household gods, and things pertaining thereto are kept. It is also much used as a mat to sit upon when performing the daily rites. Mention is made in the Smritis (the teaching of the Sages) of the purity acquired by wearing deer skin, and there are several injunctions on the matter in the laws of Manu. For example:

> *"Let the students in theology wear for their mantles the hides of black antelopes, of common deer or of goats, with lower vests of woven sana of cshuma and of wool, in the direct order of their classes." (Mann, ii. 41.)*

The piece of deer skin is worn with the thread for several months, when it is taken off, with some short ceremony, at a temple.. When the father puts on the true cord, he repeats the yajnopavitam mantram, the novice saying it after him. This mantram is as follows:

"This most hallowed yajnopavitam,
In former times with Brahma born,
Author of longevity; wear it, it is pure,
May this yajnopavitam become my strength and glory."

As the new and true cord is put on, the imitation one which was first used is taken off. This completes the investiture, and the father at once proceeds to teach the novice the gayatri prayer. This is done with great care and secrecy. A cloth is thrown over the heads of both father and son, and, the sacred words are whispered into the ears, in as low a whisper as possible, so that the holy words may not fall upon the ears of any uninitiated person. The upanayanam is now complete, and the lad is now a true dvija, duly entitled to read the Vedas, and to perform any of the religious rites of his caste.

Immediately following this investiture the youth proceeds to ask alms of those present, beginning with his mother and then his father and after wards the other relatives or friends. This act is supposed to Intimate a readiness on the part of the supplicant to provide for himself and his religious preceptor. All this takes place on the first day, but for three more days the festival is kept up, during which the novice is instructed in the morning, midday and evening prayers and in other ceremonial observances. There is always much feasting and rejoicing upon these occasions. Musicians are hired to enliven the company and friends and relatives are entertained according to the ability of the host.

A new cord must be put on every year on the occasion of a certain festival. This festival is called sravanalapaurnavami—the full moon in the month of Sravana (July-August). Should the cord be broken during the year, a new one must at once be put on. If a man has a cord of five skeins, a broken thread or two does not matter; but a bachelor must have his one skein perfect, without even a single thread being broken, and a married person must have at least three perfect skeins, every thread of which must be perfect.

There are also certain kinds of defilement, as for instance touching a Pariah, that necessitate the putting on a new thread and the casting away the old one. In these days, the orthodox are not always so very particular as this, but this is the rule. Should the cord become broken, or any defilement contracted, no food can be taken until the old is replaced by a new one.

If a strict orthodox Brahmin, in passing through the bazaar, accidentally comes into contact with a Pariah, or in any other way becomes ceremonially defiled, he must get a new cord, which he cannot touch until he has bathed, and thus purified himself from the defilement. After bathing he 'takes the new cord and, dipping it into water, spreads it out on two brass or copper vessels. He then touches it with some of the pigment he uses for putting the sacred mark on his forehead. After that he walks round the vessels three times, from right to left, repeating the gayatri prayer. Then he takes the cord, skein by skein, and puts it on saying the mantram, or consecration prayer, used at the first investiture, repeating the same for each skein.

When he has thus put on the whole of the skeins, he takes off the old cord, repeating a mantram which says:

"May this old yajnopavitam become my strength and glory."

The old thread is disposed of by throwing it into a river or some other water, if there should be any at hand. The ancient lawgiver says:

"His girdle, his leather mantle, his staff, his sacrificial cord, and his ewe; he must throw into the water, when they are worn out or broken, and receive others hallowed by mystical texts." (Mann, ii. 64.)

Should no river, or other suitable water be conveniently near, the old thread is rolled up and thrown on to the top of the house. This is to prevent its being trodden under foot, or in any other such way defiled. This completes the re-investiture. The defiled one is now ceremonially pure, and he can proceed to perform the daily rites which must be gone through before he can partake of food.

The sacred thread is usually worn over the left shoulder, hanging down across the body under the right arm, and, as the orthodox Hindu is not encumbered with much by way of covering for the upper part of his person, it forms a very noticeable object. On certain occasions, however, the position of the thread is changed. At the' time of performing the annual ceremony for deceased ancestors, the position is exactly reversed. It is then placed over the right shoulder and hangs down on the left side.

On certain other occasions, it is worn as a garland round the neck: whilst at others it is placed up over the ears to prevent its being defiled. Usually when saying the gayatri prayer, the thread is taken hold of by the thumb; and on reciting various mantrams it is used somewhat as a rosary—the worshipper winding it round the fingers to keep count of the number of times the mantram is repeated. The ancient law-giver Manu, makes various allusions to the sacred thread other than those quoted above. He speaks of the sinfulness of omitting the sacred investiture (xi. 63); and he lays down the rule that no one must use a sacerdotal string that has been before used by another (iv. 66).

The yajnopavitam is to the Hindu an all important thing, being the sign of the second or spiritual birth. Without his cord the Brahmin is not a Brahmin. He is nothing better than an outcaste, he cannot perform any ceremony or partake of any food, nay, he must not even swallow his own spittle. He may breathe, and that is about all he can do until the lost or defiled cord is duly replaced with all proper ceremony. A Sanyasi does not wear this cord, for he has entered the fourth or last stage of the Brahmin's life.

4

Hindu Way of Life

A Woman and the Moon

The Earth's one natural satellite, the Moon, is more than one quarter the size of Earth itself (3,474 km diameter). Because of its smaller size, the Moon's gravity is one-sixth of the Earth's gravity. While there are only two basic types of regions on the Moon's surface, there are many interesting surface features such as craters, mountain ranges, riles, and lava plains.

The structure of the Moon's interior is more difficult to study. The Moon's top layer is a rocky solid, perhaps 800 km thick. Beneath this layer is a partially molten zone. Although it is not known for certain, many lunar geologists believe the Moon may have a small iron core, even though the Moon has no magnetic field. By studying the Moon's surface and interior, geologists can learn about the Moon's geological history and its formation.

The footprints left by Apollo astronauts will last for centuries because there is no wind on the Moon. The Moon does not possess any atmosphere, so there is no weather as we are used to on Earth. Because there is no atmosphere to trap heat, the temperatures on the Moon are extreme, ranging from 100ºC at noon to-173ºC at night.

How Do the Phases Get Their Names?

When the Moon appears smaller than a quarter, we call it a crescent. When the Moon appears larger than a quarter, we call it gibbous. When the moon is getting bigger (phases new to full) it is waxing. When it is getting smaller (phases Full to New) it is waning.

The Moon, with its emotional sensitivity, has traditionally been viewed as feminine. The Sun, with its charisma and force, has been seen as masculine.

The moon is the celestial body which is the most visible in the nightly sky. The moon dominates the nightly sky and can be seen far more clearly than any other planet or star.

The phases of the moon are one of humankind's best and most ancient means of measuring time. "Moon," "menses," and "month" all share the same root, meaning to measure.

The most visible symbol of feminine energy in the solar system is the Moon. Because it reflects the light from the sun, the pale orbit in the night sky has come to represent a woman's reflective nature, her internalized response to the world.

Earth's moon has been personified as female and has been perceived as a Symbol of the feminine principle since time immemorial. An increasing number of contemporary people are beginning to appreciate the profound truth and the high spiritual validity of this universal perception as they rediscover and reimplement ancient spiritual traditions which honour women and embody distinctly feminine spiritual values.

The coincidence of the menstrual cycle with that of the moon is a physical actuality structuring human life and a curiosity that has been observed with wonder. It is in fact likely that the fundamental notion of life-structuring relationship between the heavenly world and that of man was derived from this realization, both in experience and in thought, of the force of the lunar cycle.

The Lunar calendar year is divided into 12 parts, each of which is called a month. But the word month has other meanings. Several kinds of months are measured by the moon's motion. At one point in the moon's path, it is closest to the earth. This point is called the perigee. The time the moon takes to revolve from one perigee to the next is an anomalistic month. This period averages 27 days, 13 hours, 18 minutes, and 33.1 seconds. If the moon were looked at from a distant star it would seem to make a complete revolution around the earth in 27 days, 7 hours, 43 minutes, and 11.5 seconds. This period is a sidereal month. The proper lunar month, which is called the synodical month, is the period between one new moon and the next, an average of 29 days, 12 hours, 44 minutes, and 2.8 seconds. It is note worthy to point out here that the average menstrual cycle of women is exactly the same as a sidereal month of Moon.

The great sage Tirumular has dealt in Thirumanthram about the similarity of the women and the moon. The moon has Four phases. Likewise a woman has also four phases of a manstural cycle. The influence of the moon on human life is very high.

The moon influences the mind and the sensations of every human. The moon also influences the physiology of a woman very much. The birth star, the moon sign and the thithi of a girl's birth have greater influence on the important physiological cycles of a woman. The day of her attaining maturity, the day of her conception and the day of her menopause can be predetermined with these.

The moon takes a raising trend from new moon to full moon and a falling trend from full moon to new moon. The Vedic scriptures term the first trend as the Suklapaksa and the second trend as the Krishna paksa.

The full moon occurs on the 14th day of the new moon and a lady is prone to conception on the 14th day of her menstrual period.

What is Menstruation?: Every month each ovary releases an ovum, which reaches the uterus via fallopian tubes. During this period, a soft and spongy lining develops in the uterus known as endometrium, which is the base of fertile ovum and develops as fetus. Unfertile ovum cannot stay there. The endometrium comes out in the form of a red liquid, for 3 to 5 days. This only is known as menstruation.

Every woman's cycle is approximately 28 days long.

The menstrual cycle is regulated by the hormones estrogen and progesterone. It has 4 phases.

1. *Menstruation:* The shedding of about three quarters of the endometrial lining of the uterus lasting 4-5 days triggered by a fall in the estrogen and progesterone levels
2. *Proliferative Phase:* Endometrium regenerating. This phase lasts around 10 to 13 days and ends with ovulation. During this time progesterone levels remain stable and low while estrogen increases. When estrogen levels reach first peak it sets off a chain of hormonal responses, stimulating ovulation.
3. *Secretory Phase:* Following ovulation the follicle from which the egg is released changes structurally into the corpus luteum and secretes progesterone. Progesterone levels and estrogen levels increase over 10 to 14 days and the endometrium itself is converted to an actively secreting tissue and the glands lining its surface become filled with glycogen. Glycogen is the main carbohydrate stored in cells.
4. *Pre-menstrual Phase:* If fertilization does not occur the corpus luteum disappears and with this its influence on hormonal levels. Over the two days prior to menstruation estrogen and progesterone levels drop to their baseline level.

Have you ever wondered about the connection between your body's 28 day cycle and the cycle of the moon? Here's the theory.

The Midpoint of the Menstrual Cycle (About Day 14 in a 28-Day Cycle as the case of full moon)

The mature ovum bursts from the follicle about midway (approximately 2 weeks before onset of next menstrual period) through the menstrual cycle. This process is known as ovulation. The ovum then travels from the ovary down the fallopian tube, and into the uterus.

As the moon is raising from the new moon to full moon the Progesterone levels and estrogen levels increase till the 14th day of the menstrual cycle and the endometrium itself is converted to an actively secreting tissue and the glands lining its surface become filled with glycogen. In a woman who hasn't become pregnant, the level of progesterone peaks about a week after ovulation and then begins to drop along with the estrogen level. The flow of blood to the endometrium decreases, and its upper portion is broken down and shed during menstruation. At the same time, the corpus luteum withers. This marks krishnapaksa, the falling trend of the moon.

The influence of moon on child birth.

A woman conceived on an Amavasya (new moon day) or Pradamai (the immediate second day from new moon/or of a full moon day thithis, delivers a spastic/under developed child or mentally unsound child.

A women who conceives on her Chandraashtama day delivers a child who will be weak, suffers frequent ailments and illness and suffers with memory loss in the young age itself.

If conceived on the Ashtami or Navami thittis, then the child likely to grow with some arrogance, aggressive tendency and never learns to adjust to the surroundings/situations.

If Amavasya day also happens to be the Chandraashtama day for women who conceived on that day, and if Surya Dasa/ bhukti too runs then, the child born will, mostly be blind or with defective eye sight.

Now, for the positive side:

Human body in itself is a mystery! No one can say for sure that one has fully control and knowledge of one's own body. Yes, now you can simply do this following test and find whether you are aware of this.

Now take a deep breath. Exhale the same slowly. Check with your pointing finger taking it near to your nose, from which nostril the breathed air is coming out. From right or left nostril? or from both side nostrils? If the exhaling is from the left nostril, it is the moon's breath and if the exhaling from the right nostril, it is the sun's breath.

For Men: If the carbon is released through your right nostril then it is a good omen. When starting to do some thing new or when moving out of your home for some auspicious functions, check your breath is emitted from your right nostril. If it is from the left nostril or when the flow is from both nostrils then wait for few minutes to continue. The best timing for having sex with your partner is while the emission is from your right side nostril.

For Women: When the emission of carbon is from your left nostril, then it is the best time for you to start your activities. Also while planning for a child and having sex see that the carbon emission is through your left nostril. If emitted from both nostrils or from the right one then it is considered inauspicious.

Interesting Facts: Some other interesting facts about the coincidence of the moon and women are described here under:

The moon is the sub planet of the earth. While all other planets are rotating the sun, the moon is only rotating the earth. Consider the nature of women assuming the sun is the husband of the moon. The earth is the parent of the moon. This evidences the fact that a woman's heart is always centering around her parents wherever she is.

A moon becomes shinning and beautiful when it goes nearer to her husband Sun. Likewise a women is honoured and respected while she is along with her husband. This is the day of full moon.

A moon becomes dark and unseen when it keeps away from the Sun. Likewise a woman looses her glory and respect when she keeps away her husband. This is the day of new moon.

The Delight of Supernal

Husband and Wife

The horizontal axis of family life is manifested primarily in the mutual love between husband and wife. The bond of marriage is regarded as divinely ordained in most religious traditions. As such, it carries with it the promise of God's blessing, and should be full of love and joy.

But love is not merely a matter of unfettered emotion. Subsequent passages spell out some of the responsibilities of marriage for both the husband and wife. The husband should honour his wife, never oppress or mistreat her, and always be faithful—and the wife should do likewise.

The scriptures of all religions also distinguish between roles of the husband and wife: the husband protects and supports his wife, the head of the household yet deferring to his wife in domestic affairs. The wife is obedient to her husband, serves him with kindness, and takes primary responsibility for raising the children. While of late these traditional roles have been questioned, they have served to strengthen the bonds of family through every generation. Finally, we include several passages on the subject of the good wife.

Not those are true husband and wife that with each other [merely] consort: Truly wedded are those that in two frames, are as one light.

1. *Sikhism. Adi Granth, Var-Suhi-Ki, M.3, p. 788*

   I am He, you are She;
   I am Song, you are Verse,
   I am Heaven, you are Earth.
   We two shall here together dwell,
   becoming parents of children.

2. *Hinduism. Atharva Veda 14.2.71*

   Sweet be the glances we exchange,
   our faces showing true concord.
   Enshrine me in your heart and let
   one spirit dwell within us.
   I wrap around you this my robe
   which came to me from Manu,
   so that you may be wholly mine
   and never seek another.

3. *Hinduism. Atharva Veda 7.36-37*

   Representing heaven and earth, I have created husband and wife. This is the beginning of the world.

4. *Tenrikyo. Mikagura-uta*

In the beginning there was only the Self, one only. He desired, "May I have a wife in order to have offspring; may I have wealth in order to perform a work!"—for desire reaches this far. Even if one wishes, one cannot obtain more than this. As long as one does not attain each of these [desires], he thinks himself to be incomplete.

He found no joy; so even today, one who is all alone finds no joy. He yearned for a second. He became as large as a man and a woman locked in close embrace. This self he split into two; hence arose husband and wife. Therefore, as Yajnavalkya used to observe, "Oneself is like half of a split pea." That is why this void is filled by woman. He was united with her, and thence were born human beings.

5. *Hinduism. Brihadaranyaka Upanishad 1.4.17 and 1.4.3*

Then the Lord God said, "It is not good that the man should be alone; I will make him a helper fit for him." So out of the ground the Lord God formed every beast of the field and every bird of the air, and brought them to the man to see what he would call them; and whatever the man called every living creature, that was its name. The man gave names to all cattle, and to the birds of the air, and to every beast of the field; but for the man there was not found a helper fit for him. So the Lord God caused a deep sleep to fall upon the man, and while he slept took one of his ribs and closed up its place with flesh; and the rib which the Lord God had taken from the man he made into a woman and brought her to the man. Then the man said, "This at last is bone of my bones and flesh of my flesh; she shall be called Woman, because she was taken out of Man." Therefore a man leaves his father and his mother and cleaves to his wife, and they become one flesh.

6. *Judaism and Christianity. Bible, Genesis 2.18-24*

Brihadaranyaka Upanishad 1.4.17 and 1.4.3: This is an account of the creation of pairs from the primordial Androgyne. Cf. Prasna Upanishad 1.4-5, p. 176; Maori Tradition, p. 311a. Genesis 2.18-24: These verses give divine sanction to marriage. Jesus used them to declare that divorce was not acceptable to God—see Mark 10.2-12, p. 475.

The verse, "And Isaac brought her into his mother Sarah's tent" (Genesis 24.67), our masters have interpreted to mean that the Divine Presence came into Isaac's house along with

Rebecca. According to the secret doctrine, the supernal Mother is together with the male only when the house is in readiness and at the time the male and female are conjoined. At such time blessings are showered forth by the supernal Mother upon them.

7. *Judaism. Zohar, Genesis 101b*

The moral man finds the moral law beginning in the relation between man and woman, but ending in the vast reaches of the universe.

8. *Confucianism. Doctrine of the Mean 12*

The point at which Adam and Eve join into one body as husband and wife is also the point at which God, the subject of love, and man, the object of beauty, become one union, thus establishing the centre of goodness. Here, for the first time, the purpose of creation is accomplished. God, our Parent, is able to abide with perfected men as His children, and peacefully rest for eternity. At that time, this centre would become the object of God's eternal love, and through this, God would be stimulated with happiness for eternity. Here God's Word would be physically incarnated for the first time in human history.... However, the universe lost this centre when man fell.

9. *Unification Church. Divine Principle I.1.2.3.4*

Blessed art Thou, O Lord our God, King of the universe, who has created all things to his glory.

Blessed art Thou, O Lord our God, King of the universe, Creator of humankind.

Blessed art Thou, O Lord our God, King of the universe, who created humankind in his image, in the image of the likeness of his form, and has prepared for him from his very own person an eternal building. Blessed art Thou, O Lord, Creator of man.

May you be glad and exultant, O barren one, when her children are gathered to her with joy. Blessed art Thou, O Lord, who makes Zion joyful through her children. May Thou make joyful these beloved companions, just as Thou gladdened Thy creatures in the Garden of Eden in primordial times. Blessed art Thou, O Lord, who makes bridegroom and bride to rejoice.

Blessed art Thou, O Lord, King of the universe, who created mirth and joy, bridegroom and bride, gladness, jubilation, dancing and delight, love and brotherhood, peace and fellowship. Quickly, O Lord our God, may the sound of mirth and joy be heard in the streets of Judah and Jerusalem, the voice of bridegroom and bride, jubilant voices of bridegrooms from their canopies and youths from the feasts of song. Blessed art Thou, O Lord, who makes the bridegroom rejoice with the bride.

10. Judaism. Talmud, Ketubot 8a

Doctrine of the Mean 12: Cf. I Ching 54, p. 123. Divine Principle I.1.2.3.4: The Blessing, or holy wedding, is the chief sacrament in the Unification Church. Blessed marriages are for eternity. Cf. Divine Principle I.2.2.2, p. 429. Ketubot 8a: These six benedictions are recited at the wedding ceremony. The 'building' refers to the creation of Eve from Adam's rib, as well as the household of the family. The reference to God as the Creator of humankind denotes that marriage is God's design for the perpetuation of the human race, which began with the blessing to Adam and Eve in Eden. The last benediction connects the joy of the newlyweds with the eschatological joy at the fulfilment of God's kingdom in Jerusalem. Bridal 'canopies' are used at all Jewish weddings.

In the celestial glory there are three heavens or degrees; and in order to obtain the highest, a man must enter into this order of the priesthood (meaning the new and everlasting covenant of marriage); and if he does not, he cannot obtain it.

And again, verily I say unto you, if a man marry a wife by my word, which is my law, and by the new and everlasting covenant, and is sealed unto them by the Holy Spirit of promise, by him who is anointed, unto whom I have appointed this power and the keys of this priesthood; and it shall be said unto them—Ye shall come forth in the first resurrection... and shall inherit thrones, kingdoms, principalities, and powers, dominions, all heights and depths—then shall it be written in the Lamb's Book of Life... and shall be of full force when they are out of the world; and they shall pass by the angels, and the gods, which are set there, to their exaltation and glory in all things, as hath been sealed upon their heads, which glory shall be a fulness and a continuation of the seeds forever and ever.

11. *Church of Jesus Christ of Latter-day Saints. Doctrine and Covenants 131.1-3, 132.19*

Among His signs is that He created spouses for you among yourselves that you may console yourselves with them. He has planted affection and mercy between you.

12. *Islam. Quran 30.21*
Set me as a seal upon your heart,
as a seal upon your arm;
For love is strong as death,
jealousy is cruel as the grave.
Its flashes are flashes of fire,
a most vehement flame.
Many waters cannot quench love,
neither can floods drown it.
If a man offered for love
all the wealth of his house,
it would be utterly scorned.

13. *Judaism and Christianity. Bible, Song of Solomon 8.6-7*

Doctrine and Covenants 131.1-3, 132.19: Latter-day Saints of pure faith, who are members of the priesthood, may enter into Temple Marriage, which establishes an eternal, indissoluble bond.

Kwan-kwan go the ospreys,
On the islet in the river.
The modest, retiring, virtuous, young lady—
For our prince a good mate is she.
Here long, there short, is the duckweed,
To the left, to the right, borne about by the current.
The modest, retiring, virtuous, young lady—
Waking and sleeping, he sought her.
He sought her and found her not,
And waking and sleeping he thought about her.
Long he thought; oh! long and anxiously;
On his side, on his back, he turned, and back again.

Here long, there short, is the duckweed;
On the left, on the right, we gather it.
The modest, retiring, virtuous, young lady—
With lutes, small and large, let us give her friendly welcome.
Here long, there short, is the duckweed;
On the left, on the right, we cook and present it.
The modest, retiring, virtuous, young lady—
With bells and drums let us show delight in her.

14. Confucianism. Book of Songs, Ode 1

Book of Songs, Ode 1: This ode begins by describing a lover's anxiety as he awaits his bride, and ends with the joy of friends and family at their wedding. Many interpret the ode as describing the virtue of a bride of King Wen, as shown by her modest disposition and retiring manner. The king's anxiety and long quest to obtain his bride is often remarked. The sound of male and female ospreys answering each other at a distance alludes to the distance between the lovers; the soft duckweed gathered and presented as an offering alludes to their union. Confucius cites this ode, see Analects 3.20, p. 921, as a model of restrained pleasure, of joy not carried to extremes. Cf. Song of Solomon 3:1-5, pp. 763f.

Kaen-kwan went the axle ends of my carriage,
As I thought of the young beauty, and went to fetch her.
It was not that I was hungry or thirsty,
But I longed for one of such virtuous fame to come and be with me.
Although no good friends be with us, we will feast and be glad.
Dense is that forest in the plain,
And there sit the long-tailed pheasants.
In her proper season that well-grown lady,
With her admirable virtue, is come to instruct me.
We will feast, and I will praise her.
"I love you, and will never be weary of you."

Although I have no good spirits,
We will drink, and perhaps be satisfied.
Although I have no good viands,
We will eat, and perhaps be satisfied.
Although I have no virtue to impart to you,
We will sing and dance.
I ascend that lofty ridge,
And split the branches of the oaks for firewood.
I split the branches of the oaks for firewood
Amid the luxuriance of their leaves.
I see you whose match is seldom to be seen,
And my whole heart is satisfied.
The high hill is seen above;
The great road is easy to travel,
My four steeds advanced without stopping;
The six reins [make music] in my hands like lute-strings.
I see you, my bride,
To the comfort of my heart.

15. *Confucianism. Book of Songs, Ode 218*

The union of hearts and minds and freedom from hate I'll bring you.

Love one another as the cow loves the calf that she has borne.

Let son be loyal to father, and of one mind with his mother; let wife speak to husband words that are honey-sweet and gentle.

Let not a brother hate a brother, nor a sister hate a sister, unanimous, united in aims, speak you words with friendliness.

I will make the prayer for that concord among men at home by which the gods do not separate, nor ever hate one another.

Be not parted—growing old, taking thought, thriving together, moving under a common yoke, come speaking sweetly to one another; I'll make you have one aim and be of one mind.

Common be your water-store, common your share of food; I bind you together to a common yoke.

United, gather round the sacrificial fire like spokes around the nave of a wheel.

With your common desire I'll make you all have one aim, be of one mind, following one leader, like the gods who preserve their immortality.

Morn and eve may there be the loving heart in you.

16. *Hinduism. Atharva Veda 3.30*

Atharva Veda 3.30: This hymn sets forth the ideal of the Hindu family. Cf. Rig Veda 10.191.2-4, p. 272.

Behold the comely forms of Surya! her border-cloth and her headwear, and her garment triply parted, these the priest has sanctified.

I take your hand for good fortune, that you may attain old age with me, your husband. The solar deities—

Bhaga, Aryaman, Savitri, Purandhi— have given you to me to be mistress of my household.

Pushan, arouse her, the most blissful one; through whom a new generation will spring to life.

She, in the ardor of her love, will meet me, and I, ardently loving, will meet her

Live you two here, be not parted, enjoy the full length of life, sporting with your sons and grandsons, rejoicing in your own abode.

May Prajapati bring forth children of us, may Aryaman unite us together till old age,

Not inauspicious, enter your husband's house, be gracious to our people and animals.

Come, not with fierce looks, not harming your husband, good to animals, kind-hearted and glorious, a mother of heroes, loving the gods, pleasant, gracious to humans and to animals.

Make her, thou bounteous Indra, a good mother of sons; grant her good fortune; give her ten sons and make her husband the eleventh.

Be a queen to your father-in-law, a queen to your mother-in-law, a queen to your husband's sisters, and a queen to your husband's brothers.

May the universal Devas and Apas join our hearts together; so may Matarisvan, Dhatri, and Dveshtri unite us both.

17. *Hinduism. Rig Veda 10.85.35-47*

Rig Veda 10.85.35-47: Vv. 35-37, 42-47. This is the traditional Hindu marriage vow and blessings. The bride is Surya, daughter of the solar deity Savitri; she is the prototype of all brides. 'Her husband the eleventh' means the wife will mother her husband in his old age; 'queen' describes the wife's status as head of the household.

A man is forbidden to compel his wife to her marital duty.

18. *Judaism. Talmud, Erubin 100b*

"Your wife has rights over you," said the Prophet, according to Abu Juhaifa.

19. *Islam. Hadith of Bukhari*

Your wives are as a tilth to you: so approach your tilth when or how you will; but do some good act for your souls beforehand, and fear God.

20. *Islam. Quran 2.223*

He who loves his wife as himself; who honours her more than himself; who rears his children in the right path, and who marries them off at the proper time of their life, concerning him it is written: "And you will know that your home is at peace."

21. *Judaism. Talmud, Yebamot 62*

Do not abuse your wife. Women are sacred. If you make your wife suffer, you will die in a short time. Our grandmother, Earth, is a woman, and in abusing your wife you are abusing her. By thus abusing our grandmother, who takes care of us, by your action you will be practically killing yourself.

22. Native American Religions. A Winnebago Father's Precepts

When women are honoured, there the gods are pleased; but where they are not honoured, no sacred rite yields rewards. When the female relations live in grief, the family soon wholly perishes; but that family where they are not unhappy ever prospers.

23. Hinduism. Laws of Manu 3.56-57

From woman is man born, inside her he is conceived;
To woman man is engaged, and woman he marries.
With woman is man's companionship.
From woman originate new generations.
Should woman die, is another sought;
By woman's help is man kept in restraint.
Why revile her of whom are born great ones of the earth?

24. Sikhism. Adi Granth, Asa-ki-Var, M.1, p. 473

It is well for a man not to touch a woman. But because of the temptation to immorality, each man should have his own wife and each woman her own husband. The husband should give to his wife her conjugal rights, and likewise the wife to her husband. For the wife does not rule over her own body, but the husband does; likewise the husband does not rule over his body, but the wife does. Do not refuse one another except perhaps by agreement for a season, that you may devote yourselves to prayer; but then come together again, lest Satan tempt you through lack of self-control. I say this by way of concession, not of command....

To the unmarried and the widows I say that it is well for them to remain single as I do. But if they cannot exercise self-control, they should marry. For it is better to marry than to be aflame with passion.

25. Christianity, Bible, 1 Corinthians 7.1-9

Each one of you has ties to others, so marry them with their family's consent and give them their marriage portions decently as matrons rather than taking them on as mistresses, nor having [any secret affairs with] them as girlfriends.... That goes for any of you who worries lest he may not control his impulses; however it is better for you to discipline yourselves.

26. Islam. Quran 4.25

A virtuous wife who, after the death of her husband, constantly remains chaste even though she have no son, will reach heaven just as do men living a life of renunciation....

27. Hinduism. Laws of Manu 5.160

It floats about, that boat of cypress wood,
There in the middle of the Ho.
With his two tufts of hair falling over his forehead,
He was my mate;
And I swear that till death I will have no other.
O mother, O Heaven,
Why will you not understand me?
It floats about, that boat of cypress wood,
There by the side of the Ho.
With his two tufts of hair falling over his forehead,
He was my only one;
And I swear that till death I will not do the evil thing.
O mother, O Heaven,
Why will you not understand me?

28. Confucianism. Book of Songs, Ode 45

The possession of many wives undermines a man's moral nature.

29. Hinduism. Srimad Bhagavatam 11.3

Laws of Manu 5.160: According to Hindu tradition, a virtuous widow will remain chaste and not remarry. However noble this ethic may be, it leaves widows destitute if their relatives or society do not take on the responsibility of supporting them. Cf. Mencius I.B.5, p. 1068. Book of Songs, Ode 45: This poem was sung by Kung Chiang, the widow of the prince Kung-po of Wei. Her mother wanted to force her into a second marriage, and she protests. The Chinese have always considered the refusal of a widow to marry again to be a great virtue. Cf. I Ching 54, p. 123.

You will not be able to deal equally between your wives, however much you wish to do so.

30. Islam. Quran 4.129

Whoever has many wives will have troubles in surfeit.

He will be deceitful, he will lie, he will betray [some of them] to have them together;

It is not certain that he can have peace to pray well.

31. African Traditional Religions. Yoruba Poem (Nigeria)

Men are the protectors and maintainers of women, because God has given the one more strength than the other, and because they support them from their means. Therefore the righteous women are devoutly obedient, and guard in the husband's absence what God would have them guard.

32. Islam. Quran 4.34

All of you are guardians and are responsible for your wards. The ruler is a guardian; the man is a guardian of his family; the lady is a guardian and is responsible for her husband's house and his offspring; and so all of you are guardians and are responsible for your wards.

33. Islam. Hadith of Bukhari

In the family women's appropriate place is within; men's, without. When men and women keep their proper places they act in accord with Heaven's great norm. Among the members of the family are the dignified master and mistress whom we term father and mother. When father, mother, sons, elder and younger brothers all act in a manner suited to their various positions within the family, when husbands play their proper role and wives are truly wifely, the way of that family runs straight. It is by the proper regulation of each family that the whole world is stabilized.

34. Confucianism. I Ching 37: The Family

My dear sisters the women, you have had a hard life to live in this world, yet without you this world would not be what it is. Wakan Tanka intends that you should bear much sorrow—comfort others in time of sorrow. By your hands the family moves.

35. *Native American Religions. Sioux Tradition of the Sacred Pipe*

Quran 4.129: The Quran sanctions a man to support as many as four wives, but this was expressly a concession in time of war, when many widows and orphans needed to be supported (Quran 4.3). But it declares that monogamy is the only equitable arrangement. I Ching 37: Cf. the Five Relations as set forth in Doctrine of the Mean 20.8, p. 241.

The whole future of the race depends upon its attitude toward children; and a race which specializes in women for "menial purposes" or which believes that the contest of the sexes in the spheres of business and politics is a worthier endeavor than the creation of tomorrow's generation, is a race which is dying.

36. *Scientology. L. Ron Hubbard, Science of Survival*

Woman, before decking yourself, make yourself acceptable to your Lord, Lest He should visit not your couch, and your make-up be gone to waste. In the woman finding acceptance with her Lord, lies beauty of her make-up. Should her make-up be acceptable, shall she have love of her Lord. Let her deck herself in fear of the Lord, joy in God her perfume, Love her sustenance.

Dedicating body and mind to her Lord, let her in love to Him be united.

37. *Sikhism. Adi Granth, Var-Suhi-Ki, M.3, p. 788*

You wives, be submissive to your husbands, so that some, though they do not obey the Word, may be won without a word by the behaviour of their wives, when they see your reverent and chaste behaviour. Let not yours be the outward adorning with braiding of hair, decoration of gold, and wearing of fine clothing, but let it be the hidden person of the heart with the imperishable jewel of a gentle and quiet spirit, which in God's sight is very precious.

So once the holy women who hoped in God used to adorn themselves and were submissive to their husbands, as Sarah obeyed Abraham, calling him lord. And you are now her children if you do right and let nothing terrify you.

38. Christianity. Bible, 1 Peter 3.1-6

Be subject to one another out of reverence for Christ. Wives, be subject to your husbands, as to the Lord. For the husband is the head of the wife as Christ is the head of the church, his body, and is himself its Saviour. As the church is subject to Christ, so let wives also be subject in everything to their husbands. Husbands, love your wives, as Christ loved the church and gave himself up for her, the he might sanctify her, having cleansed her by the washing of water with the word, that he might present the church to himself in splendour, without spot or wrinkle or any such thing, that she might be holy and without blemish.

Even so husbands should love their wives as their own bodies. He who loves his wife loves himself. For no man ever hates his own flesh, but nourishes it and cherishes it, as Christ does the church, because we are members of his body. "For this reason a man shall leave his father and mother and be joined to his wife, and the two shall become one flesh." This mystery is a profound one, and I am saying that it refers to Christ and the church; however, let each one of you love his wife as himself, and let the wife see that she respects her husband.

39. Christianity. Bible, Ephesians 5.21-33

1 Peter 3.6: The matriarch Sarah, wife of Abraham, is the model for later generations of women; cf. Isaiah 55.1-2, p. 793; Hebrews 11.11, p. 754. Ephesians 5.21-33: On this metaphor of the Church as the Bride of Christ, cf. Revelation 21.1-7, pp. 1118f.; also Isaiah 62.4-5, p. 206; Exodus Rabbah, p. 286.

The husband who wedded her with sacred texts always gives happiness to his wife, both in season or out of season.

Though he may be destitute of virtue, or seek his pleasure elsewhere, or devoid of good qualities, yet a husband must be constantly revered as a god by a faithful wife. Women need perform no sacrifice, no vow, no fast; if she obeys her husband, she will for that reason alone be exalted in heaven. A faithful wife, who desires to dwell after death with her husband, must never do anything that might displease him who took her hand, whether he be alive or dead.... She who, controlling her thoughts, words, and deeds, never slights her lord, resides after death with her husband in heaven, and is called a virtuous wife.

40. Hinduism. Laws of Manu 5.153-65

Sujata, the young wife of an eldest son of a rich merchant, Anathapindika, was arrogant, did not respect others and did not listen to the instruction of her husband and his parents. Consequently, some discord arose in the family. One day the Blessed One came to visit Anathapindika and noticed this state of affairs. He called the young wife, Sujata, to Him and spoke to her kindly, saying, "Sujata, there are seven types of wives:

A wife who is pitiless, corrupt in mind,
Neglecting husband and unamiable,
Inflamed by other men, a prostitute bent on murder,
Call that wife a slayer!
A wife who would rob her husband of his gains—
Though little be the profit that he makes,
Whether by craftsmanship, or from his trade, or by the plough—
Call that wife a robber!
The slothful glutton, bent on doing nothing,
A gossip and a shrew with strident voice,
Who brings to low account her husband's zeal and industry—
Call that wife a master!
Who with loving sympathy,
Just as a mother for her only son,
For husband cares, and over his stored-up wealth keeps watch and ward—
Call that wife a mother!
Who holds her husband in the same regard
As younger sister holds the elder born,
The meek in heart, who in his every wish her husband serves—
Call that wife a sister!
And she who is as glad her lord to see
As boon companions long apart to meet,

A gracious character of gentle birth, a fond helpmate—
Call that wife a friend!
If fearless of the lash and stick, unmoved,
All things enduring, calm, and pure in heart,
She bear obedience to her husband's word, from anger free—
Call that wife a handmaid!
Now she who's called: a mistress, slayer, thief,
Who's harsh, immoral, lacking in respect, when death comes—
Will wander in the miseries of hell.
But mother, sister or companion, slave,
In precept long established and restrained, when death comes—
Will wander in the happy heaven world.
These, Sujata, are the seven kinds of wives a man may have; and which of them are you?"
"Lord," said Sujata, "let the Exalted One think of me as a handmaid from this day forth."

5

Serving God with Faith and Confidence

Regarding accepting initiation from the spiritual master, in the Eleventh Canto of Srimad-Bhagavatam, Seventeenth Chapter, verse 27, it is stated by Lord Krishna, "My dear Uddhava, the spiritual master must be accepted not only as My representative, but as My very self. He must never be considered on the same level with an ordinary human being. One should never be envious of the spiritual master, as one may be envious of an ordinary man. The spiritual master should always be seen as the representative of the Supreme Personality of Godhead, and by serving the spiritual master one is able to serve all the demigods."

Following in the Footsteps of Saintly Persons

In the Skanda Purana it is advised that a devotee follow the past acaryas and saintly persons, because by such following one can achieve the desired results, with no chance of lamenting or being baffled in his progress.

The scripture known as Brahma-yamala states as follows: "If someone wants to pose himself as a great devotee without following the authorities of the revealed scriptures, then his activities will never help him to make progress in devotional service. Instead, he will simply create disturbances for the sincere students of devotional service." Those who do not strictly follow the principles of revealed scriptures are generally called sahajiyas—those who have imagined everything to be cheap, who have their own concocted ideas, and who do not follow the scriptural injunctions. Such persons are simply creating disturbances in the discharge of devotional service.

In this connection, an objection may be raised by those who are not in devotional service and who do not care for the revealed scriptures. An example of this is seen in Buddhist philosophy. Lord Buddha appeared in the family of a high-grade kshatriya king, but his philosophy was not in accord with the Vedic conclusions and therefore was rejected. Under the patronage of a Hindu king, Maharaja Ashoka, the Buddhist religion was spread all over India and the adjoining countries. However, after the appearance of the great stalwart teacher Shankaracarya, this Buddhism was driven out beyond the borders of India.

The Buddhists or other religionists who do not care for revealed scriptures sometimes say that there are many devotees of Lord Buddha who show devotional service to Lord Buddha, and who therefore should be considered devotees. In answer to this argument, Rupa Goswami says that the followers of Buddha cannot be accepted as devotees. Although Lord Buddha is accepted as an incarnation of Krishna, the followers of such incarnations are not very advanced in their knowledge of the Vedas. To study the Vedas means to come to the conclusion of the supremacy of the Personality of Godhead. Therefore any religious principle which denies the supremacy of the Personality of Godhead is not accepted and is called atheism. Atheism means defying the authority of the Vedas and decrying the great acaryas who teach Vedic scriptures for the benefit of the people in general.

Lord Buddha is accepted as an incarnation of Krishna in the Srimad-Bhagavatam, but in the same Srimad-Bhagavatam it is stated that Lord Buddha appeared in order to bewilder the atheistic class of men. Therefore his philosophy is meant for bewildering the atheists and should not be accepted. If someone asks, "Why should Krishna propagate atheistic principles?" The answer is that it was the desire of the Supreme Personality of Godhead to end the violence which was then being committed in the name of the Vedas. The so-called religionists were falsely using the Vedas to justify such violent acts as meat-eating, and

Lord Buddha came to lead the fallen people away from such a false interpretation of the Vedas. Also, for the atheists Lord Buddha preached atheism so that they would follow him and thus be tricked into devotional service to Lord Buddha, or Krishna.

Inquiring about Eternal Religious Principles

In the Naradiya Purana it is said, "If one is actually very serious about devotional service, then all of his purposes will be served without any delay."

Being Prepared to Give Up Everything Material for Krishna's Satisfaction. In the Padma Purana it is stated, "For one who has given up his material sense enjoyment and has accepted the principles of devotional service, the opulence of Vishnuloka [the kingdom of God] is awaiting."

Residing in a Sacred Place

In the Skanda Purana it is also said that for a person who has lived in Dwarka for six months, for one month, or even for one fortnight, there is awaiting elevation to the Vaikunthalokas and all the profits of sarupya-mukti (the privilege of having the same four-handed bodily features as Narayana).

In the Brahma Purana it is said, "The transcendental significance of Purushottama-kshetra, which is the eighty-square-mile field of Lord Jagannath, cannot be properly described. Even the demigods from higher planetary systems see the inhabitants of this Jagannath Puri as having exactly the same bodily features possessed by one in Vaikuntha. That is, the demigods see the inhabitants of Jagannath Puri as being four-handed."

When there was a meeting of great sages at Naimisharanya, Suta Goswami was reciting Srimad-Bhagavatam, and the importance of the Ganges was stated as follows: "The waters of the Ganges are always carrying the flavour of tulasi offered at the lotus feet of Sri Krishna, and as such the waters of the Ganges are ever flowing, spreading the glories of Lord Krishna. Wherever the waters of the Ganges are flowing, all will be sanctified, both externally and internally."

Accepting Only What is Necessary

In the Naradiya Purana it is directed, "One should not accept more than necessary if he is serious about discharging devotional service."

The purport is that one should not neglect following the principles of devotional service, nor should one accept the rulings of devotional service which are more than what he can easily perform. For example, it may be said that one should chant the Hare Krishna mantra at least one hundred thousand times daily on his beads. But if this is not possible, then one must minimize his chanting according to his own capacity. Generally, we recommend our disciples to chant at least sixteen rounds on their japa beads daily, and this should be completed. But if one is not even able to chant sixteen rounds, then he must make it up the next day.

He must be sure to keep his vow. If he does not strictly follow this out, then he is sure to be negligent. That is offensive in the service of the Lord. If we encourage offences, we shall not be able to make progress in devotional service. It is better if one fixes up a regulative principle according to his own ability and then follows that vow without fail. That will make him advanced in spiritual life.

Observing Fasting on Ekadasi

In the Brahma-vaivarta Purana it is said that one who observes fasting on Ekadasi day is freed from all kinds of reactions to sinful activities and advances in pious life. The basic principle is not just to fast, but to increase one's faith and love for Govinda, or Krishna. The real reason for observing fasting on Ekadasi is to minimize the demands of the body and to engage our time in the service of the Lord by chanting or performing similar service. The best thing to do on fasting days is to remember the pastimes of Govinda and to hear His holy name constantly.

Offering Respect to the Banyan Trees

In the Skanda Purana it is directed that a devotee should offer water to the tulasi plant and amalaka trees. He should offer respect to the cows and to the brahmanas and should serve the Vaishnavas by offering them respectful obeisances and meditating upon them. All of these processes will help the devotee to diminish the reactions to his past sinful activities.

Giving Up the Company of Nondevotees

Lord Chaitanya was once asked by one of His householder devotees what the general behaviour of a Vaishnava should be. In this connection, Lord Chaitanya replied that a Vaishnava should always give up the company of nondevotees. Then He explained that there are two kinds of nondevotees: one class is against the supremacy of Krishna, and another class is too materialistic. In other words, those who are after material enjoyment and those who are against the supremacy of the Lord are called avaishnava, and their company should be strictly avoided. In the Katyayana-samhita it is stated that even if one is forced to live within a cage of iron or in the midst of a blazing fire, he should accept this position rather than live with nondevotees who are through and through against the supremacy of the Lord. Similarly, in the Vishnu-rahasya there is a statement to the effect that one should prefer to embrace a snake, a tiger or an alligator rather than associate with persons who are worshippers of various demigods and who are impelled by material desire.

In the scriptures it is instructed that one may worship a certain demigod if he is desirous of achieving some material gain. For example, one is advised to worship the sun-god if he is desirous of getting rid of a diseased condition. For a beautiful wife, one may worship Uma, the wife of Lord Shiva, and for advanced education one may worship Saraswati.

Similarly, there is a list in the Srimad-Bhagavatam for worshippers of all demigods, according to different material desires. But all of these worshippers, although they appear to be very good devotees of the demigods, are still considered to be nondevotees. They cannot be accepted as devotees. The Mayavadis (impersonalists) say that one may worship any form of the Lord and that it doesn't matter, because one reaches the same destination anyway.

But it is clearly stated in the Bhagwad Gita that those who are worshippers of the demigods will ultimately reach only the planets of those demigods, while those who are devotees of the Lord Himself will be promoted to the Lord's abode, the kingdom of God. So actually these persons who are worshippers of demigods have been condemned in the Gita. It is described that due to their lusty desires they have lost their intelligence and have therefore taken to worshipping the different demigods. So in the Vishnu-rahasya these demigod worshippers are forcefully condemned by the statement that it is better to live with the most dangerous animals than to associate with these persons.

Not Accepting Unfit Disciples, Constructing Many Temples or Reading Many Books

Another stricture is that a person may have many disciples, but he should not act in such a way that he will be obliged to any of them for some particular action or some favour. And one should also not be very enthusiastic about constructing new temples, nor should one be enthusiastic about reading various types of books, save and except the ones which lead to the advancement of devotional service. Practically, if one very carefully reads Bhagwad Gita, Srimad-Bhagavatam, Teachings of Lord Chaitanya and this Nectar of Devotion, that will give him sufficient knowledge to understand the science of Krishna consciousness. One need not take the trouble of reading other books.

In the Seventh Canto of Srimad-Bhagavatam, Thirteenth Chapter, verse 8, Narada Muṇi, while discussing with Maharaja Yudhishthira the various functions of the different orders in society, especially mentions rules for the sannyasis, those persons who have renounced this material world. One who has accepted the sannyasa order of life is forbidden to accept as a disciple anyone who is not fit. A sannyasi should first of all examine whether a prospective student is sincerely seeking Krishna consciousness. If he is not, he should not be accepted. However, Lord Chaitanya's causeless mercy is such that He advised all bona fide spiritual masters to speak about Krishna consciousness everywhere. Therefore, in the line of Lord

Chaitanya even the sannyasis can speak about Krishna consciousness everywhere, and if someone is seriously inclined to become a disciple, the sannyasi always accepts him.

The one point is that without increasing the number of disciples, there is no propagation of the cult of Krishna consciousness. Therefore, sometimes even at a risk, a sannyasi in the line of Chaitanya Mahaprabhu may accept even a person who is not thoroughly fit to become a disciple. Later on, by the mercy of such a bona fide spiritual master, the disciple is gradually elevated. However, if one increases the number of disciples simply for some prestige or false honour, he will surely fall down in the matter of executing Krishna consciousness.

Similarly, a bona fide spiritual master has no business reading many books simply to show his proficiency or to get popularity by lecturing in different places. One should avoid all these things. It is also stated that a sannyasi should not be enthusiastic about constructing temples. We can see in the lives of various acaryas in the line of Sri Chaitanya Mahaprabhu that they are not very enthusiastic about constructing temples. However, if somebody comes forward to offer some service, the same reluctant acaryas will encourage the building of costly temples by such servitors. For example, Rupa Goswami was offered a favour by Maharaja Mansingh, the commander-in-chief of Emperor Akbar, and Rupa Goswami instructed him to construct a large temple for Govindaji, which cost vast amounts of money.

So a bona fide spiritual master should not personally take any responsibility for constructing temples, but if someone has money and wants to spend it in the service of Krishna, an acharya like Rupa Goswami may utilize the devotee's money to construct a nice, costly temple for the service of the Lord. Unfortunately, it happens that someone who is not fit to become a spiritual master may approach wealthy persons to contribute for temple constructions. If such money is utilized by unqualified spiritual masters for living comfortably in costly temples without actually doing any preaching work, this is not acceptable. In other words, a spiritual master needn't be very enthusiastic for constructing temple buildings simply in the name of so-called spiritual advancement.

Rather, his first and foremost activity should be to preach. In this connection, Srila Bhaktisiddhanta Saraswati Goswami Maharaja recommended that a spiritual master print books. If one has money, instead of constructing costly temples, one should spend his money for the publication of authorized books in different languages for propagating the Krishna consciousness movement.

Straightforwardness in Ordinary Dealings and Equilibrium in Loss and Gain

There is a statement in the Padma Purana: "Persons who are engaged in Krishna consciousness should never be disturbed by some material gain or loss. Even if there is some material loss, one should not be perturbed, but should always think of Krishna within himself." The purport is that every conditioned soul is always absorbed in thinking of materialistic activities; he has to free himself from such thoughts and transfer himself completely to Krishna consciousness. As we have already explained, the basic principle of Krishna consciousness is to always think of Krishna. One should not be disturbed in material loss, but, rather, should concentrate his mind upon the lotus feet of the Lord.

A devotee should not be subjected to lamentation or illusion. There is the following statement in the Padma Purana: "Within the heart of a person who is overpowered by lamentation or anger, there is no possibility of Krishna's being manifested."

The Demigods

One should not neglect to offer due respect to the demigods. One may not be a devotee of demigods, but that does not mean that he should be disrespectful to them. For example, a Vaishnava is not a devotee of Lord Shiva or Lord Brahma, but he is duty-bound to offer all respects to such highly positioned demigods. According to Vaishnava philosophy, one should offer respect even to an ant, so then what is there to speak of such exalted persons as Lord Shiva and Lord Brahma?

In the Padma Purana it is said, "Krishna, or Hari, is the master of all demigods, and therefore He is always worshipable. But this does not mean that one should not offer respect to the demigods."

Not Giving Pain to any Living Entity

This is the statement of Mahabharata: "A person who does not disturb or cause painful action in the mind of any living entity, who treats everyone just like a loving father does his children, whose heart is so pure, certainly very soon becomes favoured by the Supreme Personality of Godhead."

In so-called civilized society there is sometimes agitation against cruelty to animals, but at the same time regular slaughterhouses are always maintained. A Vaishnava is not like that. A Vaishnava can never support animal slaughter or even give pain to any living entity.

Other rules are that one should not offer foodstuff which is cooked by a non-Vaishnava, one should not worship the Deity before a nondevotee, and one should not engage himself in the worship of the Lord while seeing a nondevotee. One should begin the worship of the demigod Ganapati, who drives away all impediments in the execution of devotional service. In the Brahma-samhita it is stated that Ganapati worships the lotus feet of Lord Nrisimhadeva and in that way has become auspicious for the devotees in clearing out all impediments. Therefore, all devotees should worship Ganapati.

The Deities should not be bathed in water which has been touched by the nails or fingers. When a devotee is perspiring, he should not engage himself in worshipping the Deity. Similarly, there are many other prohibitions. For example, one should not cross or step over the flowers offered to the Deities, nor should one take a vow in the name of God. These are all different kinds of offences in the matter of executing devotional service, and one should be careful to avoid them. In the Padma Purana it is stated that even a person whose life is completely sinful will be completely protected by the Lord if he simply surrenders unto Him.

So it is accepted that one who surrenders unto the Supreme Personality of Godhead becomes free from all sinful reactions. And even when a person becomes an offender unto the Supreme Personality of Godhead Himself, he can still be delivered simply by taking shelter of the holy names of the Lord: Hare Krishna, Hare Krishna, Krishna Krishna, Hare Hare/ Hare Rama, Hare Rama, Rama Rama, Hare Hare. In other words, the chanting of Hare Krishna is beneficial for eradicating all sins, but if one becomes an offender to the holy names of the Lord, then he has no chance of being delivered.

The offences against the chanting of the holy name are as follows: (1) To blaspheme the devotees who have dedicated their lives for propagating the holy name of the Lord. (2) To consider the names of demigods like Lord Shiva or Lord Brahma to be equal to, or independent of, the name of Lord Vishnu. (Sometimes the atheistic class of men take it that any demigod is as good as the Supreme Personality of Godhead, Vishnu. But one who is a devotee knows that no demigod, however great he may be, is independently as good as the Supreme Personality of Godhead.

Therefore, if someone thinks that he can chant "Kali, Kali!" or "Durga, Durga!" and it is the same as Hare Krishna, that is the greatest offense.) (3) To disobey the orders of the spiritual master. (4) To blaspheme the Vedic literature or literature in pursuance of the Vedic version. (5) To consider the glories of chanting Hare Krishna to be imagination. (6) To give some interpretation on the holy name of the Lord. (7) To commit sinful activities on the strength of the holy name of the Lord. (It should not be taken that because by chanting the holy name of the Lord one can be freed from all kinds of sinful reaction, one may continue to act sinfully and after that chant Hare Krishna to neutralize his sins. Such a dangerous mentality is very offensive and should be avoided.) (8) To consider the chanting of Hare Krishna one of the auspicious ritualistic activities offered in the Vedas as fruitive activities (karma-kanda). (9) To instruct a faithless person about the glories of the holy name. (Anyone can take part in chanting the holy name of the Lord, but in the beginning one should not be instructed about the transcendental potency of the Lord. Those who are too sinful cannot appreciate the transcendental glories of the Lord, and therefore it is better not to instruct them in this matter.) (10) To not have complete faith in the chanting of the holy names and to maintain material attachments, even after understanding so many instructions on this matter.

Every devotee who claims to be a Vaishnava must guard against these offences in order to quickly achieve the desired success.

There are three ways of dealing with such insults. If someone is heard blaspheming by words, one should be so expert that he can defeat the opposing party by argument. If he is unable to defeat the opposing party, then the next step is that he should not just stand there meekly, but should give up his life. The third process is followed if he is unable to execute the above-mentioned two processes, and this is that one must leave the place and go away. If a devotee does not follow any of the above-mentioned three processes, he falls down from his position of devotion.

Tilaka and Tulasi Beads

In the Padma Purana there is a statement describing how a Vaishnava should decorate his body with tilaka and beads: "Persons who put tulasi beads on the neck, who mark twelve places of their bodies as Vishnu temples with Vishnu's symbolic representations [the four items held in the four hands of Lord Vishnu—conch, mace, disc and lotus], and who have vishnu-tilaka on their foreheads, are to be understood as the devotees of Lord Vishnu in this world. Their presence makes the world purified, and anywhere they remain, they make that place as good as Vaikuntha."

A similar statement is in the Skanda Purana, which says, "Persons who are decorated with tilaka or gopi-chandana [a kind of clay resembling fuller's earth which is produced in certain quarters of Vrindavana], and who mark their bodies all over with the holy names of the Lord, and on whose necks and breasts there are tulasi beads, are never approached by the Yamadutas." The Yamadutas are the constables of King Yama (the lord of death), who punishes all sinful men. Vaishnavas are never called for by such constables of Yamaraja. In the Srimad-Bhagavatam, in the narration of Ajamila's deliverance, it is said that Yamaraja gave clear instructions to his assistants not to approach the Vaishnavas. Vaishnavas are beyond the jurisdiction of Yamaraja's activities.

The Padma Purana also mentions, "A person whose body is decorated with the pulp of sandalwood, with paintings of the holy name of the Lord, is delivered from all sinful reactions, and after his death he goes directly to Krishnaloka to live in association with the Supreme Personality of Godhead."

Accepting Flower Garlands

The next instruction is that one should put on flower garlands which are offered to the Deity. In this connection, in the Eleventh Canto, Sixth Chapter, verse 46, of Srimad-Bhagavatam, Uddhava says to Krishna, "My dear Krishna, I have taken things which You have used and enjoyed, such as garlands of flowers, saintly articles, garments and ornaments, and I eat only the remnants of Your foodstuff, because I am Your menial servant.

So, therefore, I am sure that I shall not be attacked by the spell of material energy." The purport of this verse is that for any person who simply follows these rules and regulations of decorating the body with the marks of tilaka of gopi-chandana or sandalwood pulp, and who puts on the garlands which were offered to Krishna, there is no question of being conquered by the spell of material energy. At the time of death, there is no question of such a person's being called by the constables of Yamaraja. Even if one does not accept all the Vaishnava principles, but still takes the remnants of foodstuff offered to Krishna, or krishna-prasada, he will gradually become qualified to rise to the platform of a Vaishnava.

Similarly, in the Skanda Purana Lord Brahma tells Narada, "My dear Narada, anyone who puts on his neck the flower garland which was formerly used by Krishna becomes relieved from all disease and reactions to sinful activities, and gradually he is liberated from the contamination of matter."

Dancing Before the Deity

In the Dwarka-mahatmya the importance of dancing before the Deity is stated by Lord Krishna as follows: "A person who is in a jubilant spirit, who feels profound devotional ecstasy while dancing before Me, and who manifests different features of bodily expression can burn away all the accumulated sinful reactions he has stocked up for many, many thousands of years." In the same book there is a statement by Narada wherein he asserts, "From the body of any person who claps and dances before the Deity, showing manifestations of ecstasy, all the birds of sinful activities fly away upward." Just as by clapping the hands one can cause many birds to fly away, similarly the birds of all sinful activities which are sitting on the body can be made to fly away simply by dancing and clapping before the Deity of Krishna.

Bowing Down in Honour of the Deity

In the Naradiya Purana there is a statement about bowing down and offering respect to the Deity. It is said there, "A person who has performed a great ritualistic sacrifice and a person who has simply offered his respectful obeisances by bowing down before the Lord cannot be held as equals." The person who has executed many great sacrifices will attain the result of his pious activities, but when such results are finished, he has to take birth again on the earthly planet; however, the person who has once offered respects, bowing down before the Deity, will not come back to this world, because he will go directly to the abode of Krishna.

Standing Up to Receive the Lord

In the Brahmanda Purana it is said, "A person who sees the Lord's Ratha-yatra car festival and then stands up to receive the Lord can purge all kinds of sinful results from his body."

Following the Deity

A similar statement is there in the Bhavishya Purana, in which it is said, "Even if born of a lowly family, a person who follows the Ratha-yatra car when the Deities pass in front or from behind will surely be elevated to the position of achieving equal opulence with Vishnu."

Going to the Temple of Vishnu or to Places of Pilgrimage

It is stated in the Puranas, "Persons who attempt to visit the holy places of pilgrimage, like Vrindavana, Mathura or Dwarka, are actually glorified. By such travelling activities, they can pass over the desert of material existence."

In the Harı-bhakti-sudhodaya there is a statement about the benefit of visiting the temples of Lord Krishna. As we have explained previously, in Vrindavana, Mathura and Dwarka the system is that all the devotees take advantage of visiting various temples situated in those holy places. It is stated in the Hari-bhakti-sudhodaya, "Persons who are impelled by pure devotional service in Krishna consciousness and who therefore go to see the Deities of Vishnu in the temple will surely get relief from entering again into the prison house of a mother's womb." The conditioned soul forgets the trouble of living within the mother's womb during birth, but it is a very painful and terrible experience. In order to make an escape from this material condition, one is advised to visit a temple of Vishnu with devotional consciousness. Then one can very easily get out of the miserable condition of material birth.

Circumambulating the Temple of Vishnu

It is said in the Hari-bhakti-sudhodaya, "A person who is circumambulating the Deity of Vishnu can counteract the circumambulation of repeated birth and death in this material world." The conditioned soul is circumambulating through repeated births and deaths on account of his material existence, and this can be counteracted simply by circumambulating the Deity in the temple.

The Chaturmasya ceremony is observed during the four months of the rainy season in India (approximately July, August, September and October), beginning from Sravana. During these four months, saintly persons who are accustomed to travel from one place to another to propagate Krishna consciousness remain at one place, usually a holy place of pilgrimage. During these times, there are certain special rules and regulations which are strictly followed. It is stated in the Skanda Purana that during this period, if someone circumambulates the temple of Vishnu at least four times, it is understood that he has travelled all over the universe. By such circumambulation, one is understood to have seen all the holy places where the Ganges water is flowing, and by following the regulative principles of Chaturmasya one can very quickly be raised to the platform of devotional service.

Archana

Archana means worship of the Deity in the temple. By executing this process one confirms himself to be not the body but spirit soul. In the Tenth Canto, Eighty-first Chapter, verse 19, of Srimad-Bhagavatam, it is told how Sudama, an intimate friend of Krishna's, while going to the house of a brahmana, murmured to himself, "Simply by worshipping Krishna one can easily achieve all the results of heavenly opulence, liberation, supremacy over the planetary systems of the universe, all the opulences of this material world, and the mystic power of performing the yoga system.

The events leading to Sudama's murmuring this statement are as follows. Sri Krishna had ordered His friend Sudama to go to a brahmana's house and ask for some food. The brahmanas were performing a great sacrifice, and Sri Krishna told Sudama to plead with them that He and Balram were feeling hungry and needed some food. When Sudama went there, the brahmanas refused to offer anything, but the wives of the brahmanas, upon hearing that Sri Krishna wanted some foodstuff, immediately took many palatable dishes and went to offer them to Sri Krishna. In the Vishnu-rahasya, also, it is stated, "Any person within this world who is engaged in the worship of Vishnu can very easily achieve the ever-blissful kingdom of God, known as Vaikunthaloka."

Rendering Service to the Lord

It is stated in the Vishnu-rahasya, "Any person who can arrange for service to the Lord in the same way that a king is given service by his attendants is surely elevated to the abode of Krishna after death." Actually, in India the temples are just like royal palaces. They are not ordinary buildings, because the worship of Krishna should be performed in just the way that a king is worshipped in his palace. So in Vrindavana there are many hundreds of temples wherein the Deity is worshipped exactly like a king. In the Naradiya Purana it is stated, "H a person stays in the Lord's temple even for a few moments, he can surely achieve the transcendental kingdom of God."

The conclusion is that those who are rich men in society should construct beautiful temples and arrange for the worship of Vishnu, so that people may be attracted to visit such temples and thereby be offered the opportunity of dancing before the Lord or chanting the holy name of the Lord, or else of hearing the holy name of the Lord. In this way, everyone will be given the chance to elevate himself to the kingdom of God. In other words, even a common man, simply by visiting such a temple, will be able to attain the highest benedictions, not to mention the devotees who are constantly engaged in the service of the Lord in full Krishna consciousness.

In this connection, there is a statement in the Fourth Canto, Twenty-first Chapter, verse 31, of the Srimad-Bhagavatam, wherein King Prithu says to his subjects, "My dear citizens, please note that the Supreme Personality of Godhead, Hari, is actually the deliverer of all fallen, conditioned souls. No demigod can perform this act of delivering the conditioned souls, because the demigods themselves are conditioned. A conditioned soul cannot deliver another conditioned soul. Only Krishna or His bona fide representative can deliver him. The Ganges water which is flowing down from the toe of Lord Vishnu falls upon the earthly planet and other planets and thereby delivers all the conditioned sinful living entities. So what need is there to speak of the deliverance of persons who are always engaged in the service of the Lord? There is no doubt about their liberation, even if they have stocks of sinful activities from many, many births." In other words, a person who is engaged in the worship of the Deities can minimize his stock of sinful reactions coming from many,

many previous births. This process of worshipping the Deity has already been described, and one should try to follow these rules and regulations seriously.

Singing

In the Linga Purana there is a statement about glorifying and singing about the Lord. It is said there, "A brahmana who is constantly engaged in singing the glories of the Lord is surely elevated to the same planet as the Supreme Personality of Godhead. Lord Krishna appreciates this singing even more than the prayers offered by Lord Shiva."

Sankirtana

When a person loudly chants the glories of the Lord's activities, qualities, form, etc., his chanting is called sankirtana. Sankirtana also refers to the congregational chanting of the holy name of the Lord.

In the Vishnu-dharma there is a statement glorifying this process of congregational chanting: "My dear King, this word Krishna is so auspicious that anyone who chants this holy name immediately gets rid of the resultant actions of sinful activities from many, many births." That is a fact. There is the following statement in Chaitanya-charitamrita: "A person who chants the holy name of Krishna once can counteract the resultant actions of more sinful activities than he is able to perform." A sinful man can perform many, many sinful activities, but he is unable to perform so many that they cannot be wiped out by one single uttering of Krishna.

In the Seventh Canto, Ninth Chapter, verse 18, of Srimad-Bhagavatam, Maharaja Prahlada offers the following prayers to the Lord: "My dear Lord Nrisimha, if I can be elevated to the position of Your servant, then it will be possible for me to hear about Your activities. You are the supreme friend, the supreme worshipable Deity. Your pastimes are transcendental, and simply by hearing of them one can counteract all his sinful activities. Therefore, I shall not care for all those sinful activities, because simply by hearing about Your pastimes I shall get out of all the contamination of material attachment."

There are many songs about the Lord's activities. For example, there is the Brahma-samhita, sung by Lord Brahma; Narada-Pancharatra, sung by Narada Muni; and Srimad-Bhagavatam, sung by Sukhdeva Goswami. If these songs are heard by any person, hc can easily get out of the clutches of material contamination. There should be no difficulty in hearing these songs of God. They are coming down from many, many millions of years ago, and people are still taking advantage of them. So why, at this time, should one not take full advantage and thus become liberated?

In the First Canto, Fifth Chapter, verse 22, of Srimad-Bhagavatam, Narada Muni tells his disciple Vyasadeva, "My dear Vyasa, you should know that persons who are engaged in executing austerities and penances, studying the Vedas, performing big sacrifices, chanting the hymns of the Vedas, speculating on transcendental knowledge and performing charitable functions have for all their auspicious activities simply to gain a place in the association of devotees and chant the glories of the Lord." It is indicated here that chanting about and glorifying the Lord is the ultimate activity of the living entity.

Japa

Chanting a mantra or hymn softly and slowly is called japa, and chanting the same mantra loudly is called kirtana. For example, uttering the maha-mantra (Hare Krishna, Hare Krishna, Krishna Krishna, Hare Hare/ Hare Rama, Hare Rama, Rama Rama, Hare Hare) very softly, only for one's own hearing, is called japa. Chanting the same mantra loudly for being heard by all others is called kirtana.

The maha-mantra can be used for japa and kirtana also. When japa is practiced it is for the personal benefit of the chanter, but when kirtana is performed it is for the benefit of all others who may hear.

In the Padma Purana there is a statement: "For any person who is chanting the holy name either softly or loudly, the paths to liberation and even heavenly happiness are at once open."

Submission

In the Skanda Purana there is a statement about submission unto the lotus feet of the Lord. It is said there that those who are sober devotees can offer their submission to Krishna in the following three ways: (1) samprarthanatmika, very feelingly offering prayers; (2) dainyavodhika, humbly submitting oneself; (3) lalasamayi, desiring some perfectional stage. This desiring some perfectional stage in spiritual life is not sense gratification. When one realizes something of his constitutional relationship with the Supreme Personality of Godhead, he understands his original position and wants to be reinstated in this position, either as friend, servant, parent or conjugal lover of Krishna.

That is called lalasamayi, or very eagerly desiring to go to one's natural position. This lalasamayi stage of submission comes in the stage of perfect liberation, which is technically called svarupa-siddhi, when the living entity understands, by perfect spiritual advancement and revelation, his original relationship with the Lord.

In the Padma Purana there is a statement of submission in feeling by devotees praying to the Lord: "My Lord, I know that young girls have natural affection for young boys, and that young boys have natural affection for young girls. I am praying at Your lotus feet that my mind may become attracted unto You in the same spontaneous way." The example is very appropriate. When a young boy or girl sees a member of the opposite sex there is a natural attraction, without the need for any introduction. Without any training there is a natural attraction due to the sex impulse. This is a material example, but the devotee is praying that he may develop a similar spontaneous attachment for the Supreme Lord, free from any desire for profit and without any other cause. This natural attraction for the Lord is the perfectional stage of self-realization.

In the same Padma Purana there is a statement about submission in humbleness. It is stated there, "My dear Lord, there is no sinful living entity who is more of a sinner than myself. Nor is there a greater offender than myself. I am so greatly sinful and offensive that when I come to confess my sinful activities before You, I am ashamed." This is a natural position for a devotee.

As far as the conditioned soul is concerned, there is no wonder that he has some sinful activities in his past life, and this should be admitted and confessed before the Lord. As soon as this is done, the Lord excuses the sincere devotee. But that does not mean that one should take advantage of the Lord's causeless mercy and expect to be excused over and over again, while he commits the same sinful activities. Such a mentality is only for shameless persons. Here it is clearly said, "When I come to confess my sinful activities I become ashamed."

So if a person is not ashamed of his sinful activities and continues to commit the same sinful activities with the knowledge that the Lord will excuse him, that is a most nonsensical proposition. Such an idea is not accepted in any part of the Vedic literature. It is a fact that by chanting the holy name of the Lord one becomes washed clean of all sinful activities from his past life.

But that does not mean that after being washed off, one should again begin sinful activities and expect to be washed again. These are nonsensical propositions and are not admitted in devotional service. Someone may think, "For a whole week I may commit sinful activities, and for one day I will go to the temple or church and admit my sinful activities so that I can become washed off and again begin my sinning." This is most nonsensical and offensive and is not acceptable to the author of Bhakti-rasamrita-sindhu.

In the Narada-Pancharatra there is a statement of submission accompanied by the desire for perfection. The devotee says, "My dear Lord, when shall that day come when You will ask me to fan Your body, and according to Your pleasure, You will say, 'You just fan Me in this way'?" The idea in this verse is that the devotee is desiring to personally fan the body of the Supreme Personality of Godhead. That means that he is desiring to become the personal associate of the Supreme Lord. Of course, any devotee in any capacity, either as servant, friend or conjugal lover, always has direct association with the Lord.

But according to his different individual taste, a person desires for just one of these relationships.

Here the devotee is desiring to become a servant of the Lord and desires to fan the Lord, as does His internal energy, Lakshmi, the goddess of fortune. He also wishes that the Personality of Godhead will be pleased to give him directions as to how to fan. This submission with transcendental desire, or lalasamayi vijnapti, is the highest perfectional stage of spiritual realization. In the same Narada-Pancharatra, there is another expression of submission, wherein the devotee says, "My dear Lord, O lotus-eyed one, when will that day come when on the bank of the Yamuna I shall become just like a madman and continue to chant Your holy name while incessant tears flow from my eyes?"

This is another perfectional stage. Lord Chaitanya also desired that "a moment will appear unto me as twelve years of time, and the whole world will appear to me as vacant on account of not seeing You, my dear Lord." One should feelingly pray and become eager to render his particular type of service to the Lord. This is the teaching of all great devotees, especially Lord Chaitanya. In other words, one should learn how to cry for the Lord. One should learn this small technique, and he should be very eager and actually cry to become engaged in some particular type of service. This is called laulyam, and such tears are the price for the highest perfection.

If one develops this laulyam, or excessive eagerness for meeting and serving the Lord in a particular way, that is the price to enter into the kingdom of God. Otherwise, there is no material calculation for the value of the ticket by which one can enter the kingdom of God. The only price for such entrance is this laulyam lalasamayi, or desire and great eagerness.

Reciting Notable Prayers

According to great learned scholars, the whole Bhagwad Gita contains many authorized prayers, especially in the Eleventh Chapter, where Arjuna prays to the universal form of the Lord. Similarly, in the Gautamiya-ţantra all the verses are called prayers. Again, in Srimad-Bhagavatam there are hundreds of prayers to the Lord. So a devotee should select some of these prayers for his recitation. In Skanda Purana the glories of these prayers are stated as follows: "Devotees whose tongues are decorated always with prayers to Lord Krishna are always given respect even by the great saintly persons and sages, and such devotees are actually worshipable by the demigods."

Those who are less intelligent want to worship different demigods for some material gain rather than worship Krishna. But here it is stated that a devotee who is always engaged in offering prayers to the Lord is worshipable even by the demigods themselves. The pure devotees have nothing to ask from any demigod; rather, the demigods are anxious to offer prayers to the pure devotees. In the Nrisimha Purana it is stated, "Any person who comes before the Deity of Lord Krishna and begins to chant different prayers is immediately relieved from all the reactions of sinful activities and becomes eligible, without any doubt, to enter into the Vaikunthaloka."

Partaking of Prasada

There is this specific statement in the Padma Purana: "A person who honours the prasada and regularly eats it, not exactly in front of the Deity, along with charanamrita [the water offered to the lotus feet of the Lord, which is mixed with seeds of the tulasi tree], immediately can achieve the results of pious activities which are obtained through ten thousand performances of sacrificial rites."

Drinking Charanamrita

Charanamrita is obtained in the morning while the Lord is being washed before dressing. Scented with perfumes and flowers, the water comes gliding down through His lotus feet and is collected and mixed with yogurt. In this way this charanamrita not only becomes very tastefully flavoured, but also has tremendous spiritual value.

As described in the Padma Purana, even a person who has never been able to give in charity, who has never been able to perform a great sacrifice, who has never been able to study the

Vedas, who has never been able to worship the Lord—or, in other words, even one who has never done any pious activities—will become eligible to enter into the kingdom of God if he simply drinks the charanamrita which is kept in the temple. In the temple it is the custom that the charanamrita be kept in a big pot. The devotees who come to visit and offer respects to the Deity take three drops of charanamrita very submissively and feel themselves happy in transcendental bliss.

Smelling the Incense and Flowers Offered to the Deity

In the Hari-bhakti-sudhodaya there is a statement about the incense which is offered in the temple: "When the devotees smell the good flavor of the incense which is offered to the Deity, they thus become cured of the poisonous effects of material contamination, as much as one becomes cured of a snakebite by smelling the prescribed medicinal herbs." The explanation of this verse is that there is an herb found in the jungles which expert persons know how to use to revive the consciousness of one who is bitten by a snake. Simply by smelling that herb one becomes immediately relieved of the poisonous effects of the snakebite. The same example is applicable: when a person comes to visit the temple and smells the incense offered to the Deity, he is cured at that time from all his material contamination.

Any devotee coming into the temple should always offer something to the Deity—fruit, flowers, incense, etc. If one cannot offer anything in cash, something else must be offered. In India the system is that all the ladies and gentlemen who come in the morning to visit the temple bring so many things. Even one morsel of rice or one morsel of flour can be offered. It is a regulative principle that one should not go to see a saintly person or the Deity in the temple without any offering. The offering may be very humble, or it may be priceless. Even a flower, a little fruit, a little water—whatever is possible—must be offered. So when a devotee comes to offer something to the Deity in the morning, he is sure to smell the good flavour of the incense, and then at once he will become cleansed of the poisonous effect of material existence.

It is stated in the Tantra-shastra, "If the smell of the garland which was offered to the Deity in the temple enters into a person's nostrils, immediately his bondage to sinful activities becomes cleared. And even if one has no sinful activities, still, by smelling such remnants of flowers, one can advance from Mayavadi [impersonalist] to devotee." There are several instances of this, a prime one being the advancement of the four Kumaras. They were impersonalist Mayavadis, but after smelling the remnants of flowers and incense in the temple, they turned to become devotees. From the above verse it appears that the Mayavadis, or impersonalists, are more or less contaminated. They are not pure.

It is confirmed in Srimad-Bhagavatam, "One who has not washed off all reactions of sinful activities cannot be a pure devotee. A pure devotee has no more doubts about the supremacy of the Personality of Godhead, and thus he engages himself in Krishna consciousness and devotional service." A similar statement is in the Agastya-samhita: just to purify the impurities of our nostrils, we should try to smell the remnants of flowers offered to Krishna in the temple.

Touching the Deity

In the Vishnu-dharmottara there is a statement about touching the lotus feet of the Lord. It is said, "Only a person who is initiated as a Vaishnava and is executing devotional service in Krishna consciousness has the right to touch the body of the Deity." In India there was agitation during Gandhi's political movement because the lowborn classes of men like street-sweepers and chandalas are prohibited, according to the Vedic system, from entering the temple. Due to their unclean habits they are prohibited, but at the same time they are given other facilities so they may be elevated to the highest grade of devotional service by association with pure devotees. A man born in any family is not barred, but he must be cleansed. That cleansing process must be adopted. Gandhi wanted to make them clean simply by stamping them with a fictitious name, harijana ("children of God"), and so there was a great tug-of-war between the temple owners and Gandhi's followers.

But anyway, the present law is the law of all scripture—that if anyone is purified he can enter into the temple. Actually, that is the position. Only one who is properly initiated, who is properly following the rules and regulations, can enter, and touch the Deity—not all. And one who touches the body of the Deity, following such regulative principles, is immediately delivered from the contamination of material sins, and all of his desires become fulfilled without delay.

Seeing the Deity

In the Varaha Purana there is a statement praising the seeing of the Deity of Sri Krishna in the temple. A devotee says there, "My dear Vasundhara, any person who goes to Vrindavana and sees the Deity of Govindadeva is free from the courthouse of Yamaraja and is allowed to enter into the highest planetary system, in which reside the demigods." This means that even an ordinary person who goes to Vrindavana out of inquisitiveness and by chance sees the temple, especially that of Govindadeva, even if he is not elevated to the spiritual kingdom, is still assured promotion to the higher planetary systems. This means that simply by visiting the Deity of Govinda in Vrindavana one becomes highly elevated in pious life.

Observing Arati and Celebrations of the Lord

In the Skanda Purana there is the following description of the result of seeing arati (worship) of the Deity: "If someone sees the face of the Lord while arati is going on, he can be relieved of all sinful reactions coming from many, many thousands and millions of years past. He is even excused from the killing of a brahmana or similar prohibited activities.

As we have already explained, there are different ceremonies to be observed, such as the birthday of Krishna, the birthday of Lord Ramachandra, the birthday of some prominent Vaishnavas, the ceremony of Jhulana-yatra with the Lord sitting on a swing, and Dola-yatra (the Lord's activities in the month of March). In all festivals the Lord is seated on a car, and the car moves through different streets of the city so that people may take advantage of visiting the Lord. In the Bhavishya Purana it is said, "In such a ceremony, if even a candala [dog-eater], simply out of curiosity, sees the Lord on the cart, he becomes counted as one of the associates of Vishnu."

In the Agni Purana it is stated, "Any person who in gladness sees the worship of the Deity in the temple will obtain the results of kriya-yoga which are described in the Pancharatra scripture." Kriya-yoga is a system of practice much like practical devotional service, but it is especially meant for the mystic yogis. In other words, by this gradual process the mystic yogis are eventually elevated to the devotional service of the Lord.

In the Fourth Canto of Srimad-Bhagavatam, Twenty-ninth Chapter, verse 40, the importance of hearing of the pastimes of the Lord is stated by Sukhdeva Goswami to Maharaja Parikshit: "My dear King, one should stay at a place where the great acaryas [holy teachers] speak about the transcendental activities of the Lord, and one should give aural reception to the nectarean river flowing from the moonlike faces of such great personalities. If someone eagerly continues to hear such transcendental sounds, then certainly he will become freed from all material hunger, thirst, fear and lamentation, as well as all illusions of material existence."

Sri Chaitanya Mahaprabhu also recommended this process of hearing as a means of self-realization in the present age of Kali. In this age it is very difficult to follow thoroughly the regulative principles and studies of the Vedas which were formerly recommended. However, if one gives aural reception to the sound vibrated by great devotees and acaryas, that alone will give him relief from all material contamination. Therefore it is the recommendation of Chaitanya Mahaprabhu that one should simply hear from authorities who are actually devotees of the Lord. Hearing from professional men will not help. If we hear from those who are actually self-realized, then the nectarean rivers, like those which are flowing on the moon planet, will flow into our ears. This is the metaphor used in the above verse.

As stated in Bhagwad Gita, "A materialistic person can give up his material hankerings only by becoming situated in Krishna consciousness." Unless one finds a superior engagement, he will not be able to give up his inferior engagement. In the material world everyone is engaged in the illusory activities of the inferior energy, but when one is given the opportunity to relish the activities of the superior energy performed by Krishna, then he forgets all his lesser pleasures.

When Krishna speaks on the Battlefield of Kurukshetra, to the materialistic person it appears that this is simply talk between two friends, but actually it is a river of nectar flowing down from the mouth of Sri Krishna. Arjuna gave aural reception to such vibrations, and thus he became freed from all the illusions of material problems. In the Twelfth Canto of Srimad-Bhagavatam, Third Chapter, verse 15, it is stated, "A person who desires unalloyed devotional service to Lord Krishna, who is praised by transcendental sound vibrations, should always hear about His glorification and transcendental qualities. This will surely kill all kinds of inauspiciousness in the heart."

Expecting the Lord's Mercy

In the Tenth Canto, Fourteenth Chapter, verse 8, it is said, "My dear Lord, any person who is constantly awaiting Your causeless mercy to be bestowed upon him, and who goes on suffering the resultant actions of his past misdeeds, offering You respectful obeisances from the core of his heart, is surely eligible to become liberated, for it has become his rightful claim." This statement of Srimad-Bhagavatam should be the guide of all devotees. A devotee should not expect immediate relief from the reactions of his past misdeeds.

No conditioned soul is free from such reactionary experiences, because material existence means continued suffering or enjoying of past activities. If one has finished his material activities then there is no more birth. This is possible only when one begins Krishna conscious activities, because such activities do not produce reaction. Therefore, as soon as one becomes perfect in Krishna conscious activities, he is not going to take birth again in this material world. A devotee who is not perfectly freed from the resultant actions should therefore continue to act in Krishna consciousness seriously, even though there may be so many impediments.

When such impediments arise he should simply think of Krishna and expect His mercy. That is the only solace. If the devotee passes his days in that spirit, it is certain that he is going to be promoted to the abode of the Lord. By such activities, he earns his claim to enter into the kingdom of God. The exact word used in this verse is daya-bhak. Daya-bhak refers to a son's becoming the lawful inheritor of the property of the father. In a similar way, a pure devotee who is prepared to undergo all kinds of tribulations in executing Krishna conscious duties becomes lawfully qualified to enter into the transcendental abode.

Remembrance

Some way or other, if someone establishes in his mind his continuous relationship with Krishna, this relationship is called remembrance. About this remembrance there is a nice statement in the Vishnu Purana, where it is said, "Simply by remembering the Supreme Personality of Godhead all living entities become eligible for all kinds of auspiciousness. Therefore let me always remember the Lord, who is unborn and eternal." In the Padma Purana the same remembrance is explained as follows: "Let me offer my respectful obeisances unto the Supreme Lord Krishna, because if someone remembers Him, either at the time of death or during his span of life, he becomes freed from all sinful reactions."

Meditation

To meditate means to engage the mind in thinking of the form of the Lord, the qualities of the Lord, the activities of the Lord and the service of the Lord. Meditation does not mean anything impersonal or void. According to Vedic literature, meditation is always on the form of Vishnu. In the Nrisimha Purana there is a statement about meditation on the form of the Lord. It is said there, "Meditation focusing on the lotus feet of the Supreme Personality of Godhead has been accepted as transcendental and beyond the experience of material pain and pleasure. By such meditation, even one who is grossly miscreant can be delivered from the sinful reactions of his life." In the Vishnu-dharma there is a statement about meditation on the transcendental qualities of the Lord. It is said, "Persons who are constantly engaged in Krishna consciousness, and who remember the transcendental qualities of the Lord, become free from all reactions to sinful activities, and after being so cleansed they become fit to enter into the kingdom of God." In other words, no one can enter into the kingdom of God without being freed from all sinful reactions. One can avoid sinful reactions simply by remembering the Lord's form, qualities, pastimes, etc. In the Padma Purana there is a statement about remembering the activities of the Lord: "A person who is always engaged in meditation on the sweet pastimes and wonderful activities of the Lord surely becomes freed from all material contamination."

In some of the Puranas the evidence is given that if someone is simply meditating on devotional activities, he has achieved the desired result and has seen face to face the Supreme Personality of Godhead. In this connection, there is a story in the Brahma-vaivarta Purana that in the city of Pratishthanapura in South India there was once a brahmana who was not very well-to-do, but who was nevertheless satisfied in himself, thinking that it was because of his past misdeeds and by the desire of Krishna that he did not get sufficient money and opulence.

So he was not at all sorry for his poor material position, and he used to live very peacefully. He was very openhearted, and sometimes he went to hear some lectures delivered by great realized souls. At one such meeting, while he was very faithfully hearing about Vaishnava activities, he was informed that these activities can be performed even by meditation. In other words, if a person is unable to actually perform Vaishnava activities physically, he can meditate upon the Vaishnava activities and thereby acquire all of the same results. Because the brahmana was not very well-to-do financially, he decided that he would simply meditate on grand, royal devotional activities, and he began this business thus:

> *Sometimes he would take his bath in the River Godavari. After taking his bath he would sit in a secluded place on the bank of the river, and by practicing the yoga*

exercises of pranayama, the usual breathing exercises, he would concentrate his mind. These breathing exercises are meant to mechanically fix the mind upon a particular subject. That is the result of the breathing exercises and also of the different sitting postures of yoga. Formerly, even quite ordinary persons used to know how to fix the mind upon the remembrance of the Lord, and so the brahmana was doing this. When he had fixed the form of the Lord in his mind, he began to imagine in his meditations that he was dressing the Lord very nicely in costly clothing, with ornaments, helmets and other paraphernalia. Then he offered his respectful obeisances by bowing down before the Lord. After finishing the dressing, he began to imagine that he was cleaning the temple very nicely. After cleansing the temple, he imagined that he had many water jugs made of gold and silver, and he took all those jugs to the river and filled them with the holy water.

Not only did he collect water from the Godavari, but he collected from the Ganges, Yamuna, Narmada and Kaveri. Generally a Vaishnava, while worshipping the Lord, collects water from all these rivers by mantra chanting. This brahmana, instead of chanting some mantra, imagined that he was physically securing water from all these rivers in golden and silver waterpots. Then he collected all kinds of paraphernalia for worship—flowers, fruits, incense and sandalwood pulp. He collected everything to place before the Deity. All these waters, flowers and scented articles were then very nicely offered to the Deities to Their satisfaction. Then he offered arati, and with the regulative principles he finished all these activities in the correct worshipping method.

He would daily execute similar performances as his routine work, and he continued to do so for many, many years. Then one day the brahmana imagined in his meditations that he had prepared some sweet rice with milk and sugar and offered the preparation to the Deity. However, he was not very satisfied with the offering because the sweet rice had been prepared recently and it was still very hot. (This preparation, sweet rice, should not be taken hot. The cooler the sweet rice, the better its taste.) So because the sweet rice had been prepared by the brahmana very recently, he wanted to touch it so that he could know whether it was fit for eating by the Lord. As soon as he touched the sweet rice pot with his finger, he immediately was burnt by the heat of the pot. In this way, his meditation broke. Now, when he looked at his finger, he saw that it was burnt, and he was wondering in astonishment how this could have happened. Because he was simply meditating on touching the hot sweet rice, he never thought that his finger would actually become burnt.

While he was thinking like this, in Vaikuntha Lord Narayana, seated with the goddess of fortune, Lakshmi, began to smile humorously. On seeing this smiling of the Lord, all the goddesses of fortune attending the Lord became very curious and asked Lord Narayana why He was smiling. The Lord, however, did not reply to their inquisitiveness, but instead immediately sent for the brahmana.

An airplane sent from Vaikuntha immediately brought the brahmana into Lord Narayana's presence. When the brahmana was thus present before the Lord and the goddesses of fortune, the Lord explained the whole story. The brahmana was then fortunate enough to get an eternal place in Vaikuntha in the association of the Lord and His Lakshmis. This shows how the Lord is all-pervading, in spite of His being locally situated in His abode. Although the Lord was present in Vaikuntha, He was present also in the heart of the brahmana when he was meditating on the worshipping process. Thus, we can understand that things offered by the devotees even in meditation are accepted by the Lord, and they help one achieve the desired result.

In the Naradiya Purana there is a statement of how this servitorship is transcendental. It is said there that a person who is constantly engaged in devotional service by his body, mind and words, or even a person who is not practically engaged but is simply desiring to be so, is considered to be liberated.

Devotional Service in Friendship

Devotional service in friendship can be divided into two categories: the first is to act as the confidential servant of the Lord, and the other is to act as the well-wisher of the Lord. The devotee who has confidence in devotional service to the Lord systematically follows the rules and regulations, with the faith that lie will achieve the platform of transcendental life. The second type of devotional friendship is to become a well-wisher of the Supreme Personality of Godhead. In Bhagwad Gita it is said that the Lord accepts a preacher as the most dear servant. Anyone who is preaching the confidential message of the Gita to the people in general is so dear to Krishna that no one can be equal to him in human society.

In the Mahabharata, Draupadi says, "My dear Govinda, Your promise is that Your devotee can never be vanquished. I believe in that statement, and therefore in all kinds of tribulations I simply remember Your promise, and thus I live." The purport is that Draupadi and her five husbands, the Pandavas, were put into severe tribulations by their cousin-brother Duryodhana, as well as by others. The tribulations were so severe that even Bhishmadeva, who was both a lifelong brahmachari and a great warrior, would sometimes shed tears thinking of them. He was always surprised that although the Pandavas were so righteous and Draupadi was practically the goddess of fortune, and although Krishna was their friend, still they had to undergo such severe tribulations. Though their tribulations were not ordinary, Draupadi was not discouraged. She knew that because Krishna was their friend, ultimately they would be saved.

A similar statement is there in the Eleventh Canto of Srimad-Bhagavatam, Second Chapter, verse 53, where Havi, the son of King Rishabha, addresses Maharaja Nimi: "My dear King, a person who never deviates even for a moment from engagement in service at the lotus feet of the Supreme Person (engagement which is sought even by great demigods like Indra), with firm conviction that there is nothing more worshipable or desirable than this, is called the first-class devotee."

Sri Rupa Goswami says that a neophyte devotee who has simply developed a slight love of Godhead is certainly a prospective candidate for devotional service. When he becomes firmly fixed in such devotional service, that assured status becomes a confidential part of his devotional service. Sometimes it is found that a pure devotee lies down in the temple of the Lord in order to serve Him as a confidential friend. Such friendly behaviour of a devotee may be accepted as raganuga, or spontaneous. Although, according to regulative principles, no one can lie down in the temple of the Supreme Personality of Godhead, this spontaneous love of Godhead may be grouped under devotional service in friendship.

Surrendering Everything to the Lord

Regarding complete self-surrender, there is a nice description in the Eleventh Canto of Srimad-Bhagavatam, Twenty-ninth Chapter, verse 34, where the Lord says, "A person who has completely surrendered unto Me and has completely given up all other activities is protected by Me personally, both in this life and in the next. In other words, I wish to help him become more and more advanced in spiritual life. Such a person is to be understood as having already achieved sarshti [having equal opulences with the Supreme]." It is also confirmed in Bhagwad Gita that as soon as a person surrenders unto the lotus feet of Krishna, Krishna takes charge of him and gives him a guarantee of protection from all sinful reactions. He also instructs from within, so that the devotee may very quickly make advancement toward spiritual perfection.

This self-surrender is called atma-nivedana. According to different authorities, self is differently defined. Self is sometimes considered to refer to the spirit self, or soul, and self is sometimes considered to refer to the mind or to the body. Full self-surrender, therefore, means not only surrendering one's self as spirit soul, but also surrendering one's mind and body to the service of the Lord. Srila Bhaktivinoda Thakura has sung a nice song in this connection.

While offering himself as a fully surrendered soul, he said, "My mind, my household affairs, my body, whatever is in my possession, my dear Lord, I offer to You for Your service. Now You can do with them as You like. You are the supreme possessor of everything, so if You like You can kill me, or if You like You can give me protection. All authority belongs to You. I have nothing to claim as my own." Sri Yamunacharya, in his prayers to the Lord, has expressed a similar idea in the following words: "My dear Lord, I may be living within some body as a human being or as a demigod, but whatever mode of life, I do not mind, because these bodies are simply by-products of the three modes of material nature, and I, who am in possession of these bodies, am surrendering myself unto You."

In the Hari-bhakti-viveka, there is a statement regarding how one can offer his body in self-surrender. There the devotee says, "My dear Lord, as a sold animal has no need to think about his maintenance and sustenance, so, because I have given up my body and soul unto You, I am no longer concerned with my maintenance and sustenance." In other words, one should not bother about his personal or family maintenance or sustenance. If one is actually surrendered in body and soul, he should always remember that his only concern is to be engaged in the service of the Lord.

Srila Rupa Goswami says that devotional service in friendship and devotional service in self-surrender are two difficult processes. Therefore such relationships with the Lord can very rarely be seen. Only for the advanced devotees are these two processes easily executed. The purport is that it is very rare to see surrender which is mixed with sincere ecstatic devotion. One must give himself completely to the will of the Lord.

Offering a Favourite Article

In the Eleventh Canto of Srimad-Bhagavatam, Eleventh Chapter, verse 41, Lord Krishna tells Uddhava, "My dear friend, if someone offers Me the best thing in his possession, or anything which is very pleasing to him, he will be eternally benefited."

6

Karma in the Dharma-based Religions

Hinduism

Karma came into being in Hinduism, based on the Vedas and Upanishads. One of the first and most dramatic illustrations of Karma can be found in the great Hindu epic, the Mahabharata. The original Hindu concept of karma was later enhanced by several other movements within the religion, most notably Vedanta, Yoga, and Tantra.

Hinduism sees karma as immutable law with involuntary and voluntary acts being part of a more intricate system of cause and effect that is often not comprehensible to one bound by karma. It is the goal of the Hindu, as expressed succinctly in the Bhagavad Gita, to embrace a 'sattvic' lifestyle and thus avoid creating more karma (karma is not qualified as good or bad). By ceasing to create more karma, the jivatma or individual soul is able to move closer to Moksha, or liberation.

To the Hindu, karma is the law of the phenomenal cosmos that is part and parcel of living within the dimensions of time and space. All actions, thoughts, vibrations of any sort, are governed by a law that demands perfect rebound.

So all jivatmas (individual souls) must experience karma if they live and experience the phenomenal universe. To escape the cycle of life, death and rebirth, one must exhaust one's karma and realize one's true Self as the highest truth of Oneness that is Brahman (or for dvaitists (dualists) bliss with the Supreme Godhead). In Hinduism, karma is of three kinds:

Prarabadha Karma

This karma is unchangeable within the scope of one life, since it is the 'setup' for the life in question. It is the karma of one's past lives. After death, the atma leaves the body, as the casting off of old vestments, and carries with it the samskaras (impressions) of the past life of thoughts and actions and events. These samskaras manifest themselves in the unchangeable situation into which one is born and certain key events in one's life. These include one's time of death (seen as governed by an allotment from birth of the total number of one's breaths for that life), one's economic status, one's family (or lack of family), one's body type and look: essentially, the setting of one's birth, the initial base.

Samchita Karma

The samskaras that one inherits from the last lives create one's personality, inclinations, talents, the things that make up one's persona. One's likings, abilities, attitudes and inclinations are based on the thoughts and actions of past lives. One's samchita karma is somewhat alterable through practice and effort towards change. This might be seen through the Hindu system of Yoga and the dynamic of the gunas. An example would be someone who, through meditation, slowly evolved into a more stable personality.

Agami Karma

Agami karma is the karma of the present life over which the soul has complete control. Through it one creates one's karma in the present for the future of the current life and in life-times to come. The Hindu cannot say, sometimes, if a major event in life is the doing of Prarabadha or Agami Karma. The idea of "bad things happening to good people" is seen by the Hindu as a result of Prarabadha Karma, more simply understood as karma from a past life. In Hinduism, karma works within a cyclical framework that sees the phenomenal universe being created and eventually dissolving back into itself, back into realization that it was nothing other than Maya imposed on the truth of Brahman. So Karma will eventually be worked out.

Karma does allow for anirudh (Divine Grace). Through exceeding devotion and love of God, the Hindu believes one can be helped to speed through Karma phal (Karmic fruit). By developing 'vairagya' or 'detachment' from the fruits of one's karma, as Lord Krishna most famously summarized, one can transcend karma and be liberated. One is aided by love of God. All the Yogas of Hinduism seek to transcend karma through different means of realization.

Buddhism

In Buddhism, only intentional actions are karmic "acts of will". The 'Law of Karma' refers to "cause and effect", but Karma literally means "action"-often indicating intent or cause. Accompanying this usually is a separate tenet called Vipaka, meaning result or effect. The re-action or effect can itself also influence an action, and in this way, the chain of causation continues ad infinitum. When Buddhists talk about karma, they are normally referring to karma/action that is 'tainted' with ignorance-karma that continues to ensure that the being remains in the everlasting cycle of samsara.

This samsaric karma comes in two 'flavours'-'good' karma, which leads to positive/pleasurable experiences, like high rebirth (as a deva, asura, or human), and bad karma which leads to suffering and low rebirth (as a hell-sufferer, as a preta, or as an animal). There is also a completely different type of karma that is neither good nor bad, but liberating. This karma allows for the individual to break the uncontrolled cycle of rebirth which always implies suffering, and thereby leave samsara to permanently enter Nirvana.

The Buddhist sutras explain that in order to generate liberating karma, we must first develop incredibly powerful concentration, and proper insight into the (un)reality of samsara. This concentration is akin to the states of mind required to be reborn in the Deva realm, and in itself depends upon a very deep training in ethical self-discipline.

This differentiation between good karma and liberating karma has been used by some scholars to argue that the development of Tantra depended upon Buddhist ideas and philosophies. Understanding the universal law of Karma provides order to a beginningless and endless universe. Alongside this view is the related notion of Buddhist rebirth-sometimes understood to be the same thing as reincarnation-which has its roots in the principle of Karma.

Jainism

Jains believe that karma is a form of matter. Mahavira described karma as "clay particles". Jains do not believe in "good karma" or "bad karma"; they try to avoid all karma.

Western Interpretation

Although, the Karma can not be proven as easily as the law of gravity, millions of people believe in it and is a part of many cultures and psyche of groups of people. Even more people, without a religious background, especially in a western culture, come to the conviction of the existence of Karma. For some, karma is a more reasonable concept than eternal damnation for the wicked. Spirituality or a belief that virtue is rewarded and sin creates suffering; eventually leads to a belief in Karma. With any law, it is better to obey the law even if one doesn't believe in it and ignorance does not grant immunability or one learns the hard way. A common expression heard is 'It's unfair' by those ignorant of Karma. A study in Karma reveals that is every thing is fair but again that depends one's conviction.

According to Karma, performance of positive action results with the reaction of a good conditioning in one's experience, whereas a negative action results in a reaction of a bad response. This may be an immediate result following the act, or a delayed result occurring either in the present life or the next. Thus, meritorious acts may create rebirth into a higher station, such as a superior human being or a godlike being, while evil acts result in rebirth as a human living in less desirable circumstances, or as a lower animal. While the action of karma may be compared with the Western notions of sin and judgment by God or gods, Karma is held to operate as an inherent principle of the Universe without the intervention of any supernatural being.

Most teachings say that for common mortals, having an involvement with Karma is an unavoidable part of day-to-day living. However, in light of the Hindu philosophical school of Vedanta, as well as Gautama Buddha's teachings, one is advised to either avoid, control or become mindful of the effects of desires and aversions as a way to moderate or change one's karma (or, more accurately, one's karmic results). An interesting similarity is found between the concept of the 'law of Karma' and the law in Westerns physics which says 'action is reaction'. One could argue that the law of Karma is the mental equivalent of this law in physics.

New Age and Theosophy

The idea of karma was popularized in the west through the work of the Theosophical Society. Kardecist and Western New Age reinterpretations of karma frequently cast it as a sort of luck which is associated with virtue: if one does good or spiritually valuable acts, one deserves and can expect good luck; contrariwise, if one does harmful things, one can expect bad luck or unfortunate happenings. In this conception, karma is affiliated with the Neopagan *law of return* or *Threefold Law,* the idea that the beneficial or harmful effects one has on the world will return to oneself.

Moksha (Sanskrit: *liberation*) or mukti (Sanskrit: *release*) refers, in general, to liberation from the cycle of death and rebirth. In higher Hindu philosophy, it is seen as a transcendence of phenomenal being, of any sense of consciousness of time, space and causation (Karma). It is not seen as a soteriological goal in the same salvific sense as, say, in a Judeo-Christian context, but signifies dissolution of the sense of "I", or ego, and the overall breakdown of *nama-roopa* (name-form). It is, in Hinduism, viewed as analogous to nirvana, though Buddhist thought tends to differ with even the Advaita Vedantist reading of liberation. Jainism also believes in moksha.

Hinduism, in support of the idea of moksha, posits the idea of atman and Brahman. A common mistake is to view them, both spoken of as *Self,* as a monist being of sorts, something possessing substances. In actuality, Hindu scripture like the

Upanishads and Bhagvad Gita, and especially the non-dual Hindu school of Advaita Vedanta, say that the Self or Super-Soul is beyond being and non-being, beyond any sense of tangibility and comprehension. Moksha is seen as a final release from one's worldly conception of self, the loosening of the shackle of experiential duality and a re-establishment in one's own fundamental nature, though the nature is seen as ineffable and beyond sensation.

Means to Achieve Moksha

There a four yogas (unions) or margs (paths) for the attainment of moksha. They are the ways of selfless work, of self-dissolving love, of absolute discernment, and of 'royal' meditative immersion. Different schools of Hinduism place varying emphasis on one path or other, some of the most famous being the tantric and Yogic practices developed in Hinduism. Today, the two major schools of thought are Advaita Vedanta and Bhakti branches.

1. Bhakti sees the Self as God, most often a personified monotheistic conception of Vishnu, Shiva or Devi (Mother Goddess). Unlike in Abrahamic traditions, this monotheism does not prevent a Hindu from worship of other gods, beings or teachers, as they are all seen as rays from a single source. The concept is essentially of self-dissolution in love, since the ideal nature of being is seen as that of harmony, euphony, its manifest essence being love. By immersing oneself in the love of God, one's Karmas (good or bad, regardless) slough off, one's illusions about beings decay and 'truth' is soon known and lived.
2. Vedanta finds itself split three-fold, though the dualist and modified non-dualist schools are primarily associated with the foregoing thought of Bhakti. The most famous today is Advaita Vedanta, a non-dual (*i.e.* no separation between the individual and reality/God/ etc.) perspective which often played the role of Hindu foil to contemporary Buddhist philosophy. In general, it focused on intense meditation and moral realignment,

its bedrock being the Upanishads, Brahma Sutras and the teachings of its putative founder, Adi Shankaracharya. Through discernment of the real and the unreal, as a peeling of the layers of an onion, the sadhak (practitioner) would unravel the Maya (illusion) of being and the cosmos to find nothing within, a nothingness which was paradoxically being, and transcendentally beyond both such inadequate descriptions. This was moksha, this was atman and Brahman realized as the substance and void of existential duality.

Mantra is a syllable or string of syllables, from the Sanskrit language, first found in Hinduism. Mantras may or may not conform to grammatical rules. Their use varies with the school and philosophy associated with the mantra. They are primarily used as spiritual conduits, words and vibrations that instill one-pointedness in the devotee. Other purposes have included religious ceremonies to accumulating wealth, avoiding danger, or eliminating enemies. Mantras originated in India with Vedic Hinduism and were later adopted by Buddhists and Jains, now popular in various modern forms of spiritual practice which are based on practices of these Indian religions.

Mantra Generally

Mantras (Sanskrit) have some features in common with spells, in that they are a translation of the human will or desire into a form of action. As symbols, sounds are seen to effect what they symbolize. Vocal sounds are frequently thought of as having magical powers, or even of representing the words, or speech of God.

For the authors of the Hindu scriptures of the Upanishads, the syllable P 'Aum', itself constituting a mantra, represents Brahman, the Godhead, as well as the whole of creation. Merely pronouncing this syllable is to experience the divine in a very direct way. Kukai suggests that all sounds are the voice of the Dharmakaya Buddha–*i.e.* as in Hindu Upanishads and Yoga thought, these sounds are manifestations of ultimate reality. We should not think that this is peculiar to Eastern culture, however. Words do have a mysterious power to affect us. Accepted scholarly etymology links the word with 'manas' meaning 'mind' and 'trana' for protection so that a mantra is something which protects the mind – however in practice we will see that mantra is considered to do far more than simply protect the mind.

For many cultures it is the written letters that have power– the Hebrew Kabbalah for instance. Letters can have an oracular function even. But in India special conditions applied that meant that writing was very definitely inferior to the spoken word. The Brahmins were the priestly caste of the Aryan peoples. It was they that preserved the holy writings – initially the Vedas, but later also the Upanishads. For years, they were the only ones who knew the mantras or sacred formulas that had to be chanted at important occasions. However, with the advent of egalitarian Hindu schools of Yoga, Vedanta, Tantra and Bhakti, it is now the case that intra-family and community mantras are passed on freely as part of generally practiced Hindu religion.

Such was the influence of the orthodox attitude of the elite nature of mantra knowledge, that even the Buddhists, who repudiated the whole idea of caste, and of the efficacy of the old rituals, called themselves the shravakas, that is 'the hearers'. A wise person in India was one who had "heard much". Mantras then are *sound symbols*. What they symbolize, and how they function depends on the context, and the mind of the person repeating them. Studies in sound symbolism suggest that vocal sounds have meaning whether we are aware of it or not. And indeed that there can be multiple layers of symbolism associated with each sound. So even if we do not understand them, mantras are no simply meaningless mumbo jumbo – no vocal utterance is entirely without meaning. We can look at mantra is a range of different contexts to see what they can mean in those contexts: Om may mean something quite different to a Hindu and a Tibetan Buddhist.

Tantras eventually came to see the letters as well as the sounds as representatives of the divine, it was when Buddhism travelled to China from India, that a shift in emphasis towards writing came about. China lacked a ecclesiastic language like Sanskrit, and achieved it's cultural unity by having a written language that was flexible in pronunciation but more precise in terms of the concepts that each character represented. In fact the Indians had several scripts which were all equally serviceable for writing Sanskrit.

Hence, the Chinese prized written language more highly than did the Indian Buddhists, and the writing of mantras became a spiritual practice in its own right. So that whereas Brahmins had been very strict on correct pronunciation, the Chinese, and other Far-Eastern Buddhists were more concerned with correctly writing something down. The practice of writing mantras as a spiritual practice, became very refined in Japan, and the writing in the Siddham script in which the Sanskrit of many Buddhist Sutras where written, is only seen in Japan nowadays. However, mantra-repetition written in Sanskrit, with any number of scripts, is well-known to many Hindu sects in India as well.

Mantra in Hinduism

Mantras was originally conceived in the great Hindu scriptures, known as the Vedas. Within practically all Hindu scriptures, the writing is formed in painstakingly crafted two line 'shlokas' and most mantras follow this pattern, although mantras are often found in single line or even single word combinations.

The basic mantra is P, or 'Aum', which in Hinduism is known as the 'pranava mantra,' the source of all mantras. The idea behind this is the Hindu philosophy of nama-rupa (name-form), which supposes that all ideas or entities in existence, within the phenomenological cosmos, have name and form. The basic name and form is the primordial vibration of Aum, as it is the first manifested nama-rupa of Brahman. Before existence and beyond existence, there is only One reality, Brahman, and the first manifestation of Brahman in existence is Aum. For this very reason, Aum is considered to be the most fundamental and powerful mantra, and thus is prefixed and suffixed to all Hindu prayers. While some mantras may invoke individual

Gods, the fundamental mantras, like 'Aum,' the 'Shanti Mantra,' the 'Gayatri Mantra' and others, all focus on the One reality. In the Hindu Tantras the universe is sound. The supreme (para) brings forth existence through the Word (Shabda). Creation consists of vibrations at various frequencies and amplitudes giving rise to the phenomena of the world. The purest vibrations are the varna, the imperishable letters which are revealed to us, imperfectly as the audible sounds and visible forms.

Varnas are the atoms of sound. A symbolic association was built up between letters and the elements, gods, signs of zodiac, parts of the body – letters became rich in these associations. For example in the Aitrareya-aranya-Upanishad we find:

> *"Mute consonants represent the earth, the sibilants the sky, the vowels heaven. The mute consonants represent fire, the sibilants air, the vowels the sun... The mute consonants represent the eye, the sibilants the ear, the vowels the mind"*

In effect each letter became a mantra and the language of the Vedas, Sanskrit, corresponds greatly to the nature of things. Thus Vedas come to represent reality itself. The syllable 'Om', represents the underlying unity of reality, which is Brahman

Japa

Japa was a concept of the Vedic sages that incorporates mantras as one of the main forms of worship, whose ultimate end is seen as moksha/liberation. Essentially, japa means repetition, and has become an established practice of all Hindu streams, from the various Yoga to Tantra. It involves repetition of a mantra over and over again, in cycles of auspicious numbers (in multiples of three), the most popular being 108. For this reason, Hindu malas (bead necklaces) developed, containing 108 beads and a head 'meru' bead. The devotee performing japa using his/her fingers counts each bead as he/she repeats the chosen mantra. Having reached 108 repetitions, if he/she wishes to continue another cycle of mantras, the devotee must turn the mala around without crossing the 'meru' bead and repeat.

It is said that through japa the devotee attains one-pointedness, or extreme focus, on the chosen deity or principle idea of the mantra. The vibrations of thc mantra are considered extremely important, and thus reverberations of the sound are supposed to awaken the prana or spiritual life force and even stimulate chakras according to many Hindu schools of thought.

Some Hindu Mantras

The most representative mantra of all the Hindu mantras is the famed 'Gayatri Mantra':

Om Bhur Buva Sva Tatsavitur Varenyam Bhargo Devasya Dhimahi Dhiyo Yo Nah Prachodayat

Gayatri Mantra is the most universal Hindu mantra and invokes the Brahman as the principle of knowledge and the illumination of the primordial Sun. Many Brahmins till today, in a tradition that has continued for at least 5,000 years, perform morning ablutions at the bank of a river and while holding the sacred thread given to them at adolescence, repeat the Gayatri Mantra.

Lead me from Ignorance to Truth

asato ma Sadhgamayam tamaso ma jyothir-Gamayam mrityor-ma amritham Gamayam Om santi santi santi

"Aum Asato ma sad gamaya, tamaso ma jyotir gamaya, mrityor ma amritam gamaya"

"Aum Lead me from ignorance to truth, from darkness to light, from death to immortality."

Hare Krishna Mantra: A very famous mantra is that created by Shri Chaitanya Mahaprabhu, a great Vaishnava bhakta (loving devotee) of the Hindu Lord Vishnu in the 15th century. It is beloved by most Hindus as very powerful, Vaishnavs and Shaivaites alike:

"Hare Krishna Hare Krishna Krishna Krishna Hare Hare, Hare Rama Hare Rama Rama Rama Hare Hare"

When A.C. Bhaktivedanta brought his ISKCON (International Society for Krishna Consciousness) brand of Hindu Vaishnavism to the West, he framed the "Hare Krishna" mantra as the sole mantra for moksha/liberation and his sect is now commonly known in the West by the name 'Hare Krishnas.'

Shanti Mantra : The Mantra of Peace

Aum Saha Navavatu / Saha Nau Bhunaktu / Saha Viryam Karvavahe / Tejaswi Navadhi Tamastu / Ma Vidvishavahe / Aum Shanti Shanti Shantihi / Hari Hi Aum.

(Om, May the Lord protect us, May the Lord allow us to enjoy, May we work together. May our studies be thorough and faithful. May we never misunderstand each other. Om, Peace, Peace, Peace!)

Universal Prayer

Sarveuaam Svastir Bhavatu / Sarveuam Santir Bhavatu / Sarveuam Purgam Bhavatu / Sarveuam mangalam Bhavatu (May good befall all, May there be peace for all, May all be fit for perfection, and May all experience that which is auspicious.)

Sarve Bhavantu Sukhinaha / Sarve Santu Giramayaha / Sarve Badragi Pasyantu / Ca Kascidh-dhuhkha Bhaga-Bhavet (Om, May all be happy. May all be healthy. May we all experience what is good and let no one suffer. Om, Peace, Peace, Peace!)

Thou are That

Tat Twam Asi

To list most of the main Hindu mantras would take reams of paper. Indeed, simple two-lined shlokas from holy Hindu texts like the Vedas, Upanishads, Bhagavad Gita, Yoga Sutra, even the Mahabharata and Ramayana, are considered powerful and finely honed enough to be repeated to great effect.

A very common mantra is formed by taking a deity's name and saluting it with in such a manner: 'Aum namah' or 'Aum Jai (Hail!)' or several such permutations. Common examples are 'Aum namah Shivaya' (Aum to Lord Shiva), 'Om Jai Jagadish Hare' (Salutations to the Universal God Vishnu), 'Aum Shri Ganeshaya Namah' (Aum to Shree Ganesha) and 'Jai Ma Kali.' Needless to say, there are more mantras of this sort than names of the Hindu deities. Another example of Hindu mantras is seen in the Surya Namaskara, whose practice is found recorded as far back as in the Hindu scripture of the Ramayana. As each of the twelve postures is performed, a salutatory mantra is recited to the Lord Surya, the Sun.

The Hindu Bija Mantra

In Hinduism the concept of mantra as mystical sounds was carried to its logical conclusion in 'seed' (Sanskrit *bija*) mantras that have no precise meaning but instead are thought to carry within their sounds connections to various spiritual principles and currents. For example, worship of the Mother Goddess Kali, in mantra form, is famously reduced to the powerful Bija mantras of the Shakti tradition of Hinduism:

"Aum Kreeng Kreeng Kreeng Hoong Hoong Hreeng Hreeng
Dakshina Kalike
Kreeng Kreeng Kreeng Hoong Hoong Hreeng Hreeng Swaha ||"

Of course, the most revered of all Bija mantras is Om / Aum.

The Bija mantra is part of the Hindu monist theory that while reality manifests itself as many/multiple, it is ultimately one.

Mantra in Buddhism

Buddhism, naturally following from Vedic society, also developed its own system and understanding of mantra, which while similar to that of Hinduism's, also took on its own particularities, especially according to region.

Mantra in Shingon Buddhism: Kukai advanced a general theory of language based on his analysis of two forms of Buddhist ritual language: *dharani* and *mantra*. Mantra is restricted to esoteric Buddhist practice whereas dharani is found in both esoteric and exoteric ritual. Dharanis for instance are found in the Pali Canon see below. Kukai coined the word 'shingon' (lit true word) as a Japanese translation of mantra.The word dharani, derives from a Sanskrit root dhar, which means to maintain. Ryuichi Abe suggests that it is generally understood as a mnemonic device which encapsulates the meaning of a section or chapter of a sutra. This is perhaps related to the use of verse summaries at the end of texts as in the Udana which is generally acknowledged as being in the oldest strata of the Pali Canon. Dharanis are considered to protect the one who chants them from malign influences and calamities.

Mantra is said to be derived from two roots: 'man', to think; and the action oriented (k.rt) suffix 'tra'. Thus a mantra can be considered to be a linguistic device for deepening ones thought, or in the Buddhist context for developing the enlightened mind. However, it is also true that mantras have been used as magic spells for mundane purposes such as attaining wealth.

The distinction between dharani and mantra is a difficult one to make. We can say that all mantras are dharanis but that not all dharanis are mantras. Mantras do tend to be shorter. Both tend to contain a number of unintelligible phonic fragments such as Om, or Hum which is perhaps why some people consider them to be essentially meaningless. Kukai made mantra a special class of dharani which showed that every syllable of a dharani was a manifestation of the true nature of reality – in Buddhist terms that all sound is a manifestation of shunyata or emptiness of self-nature. Thus rather than being devoid of meaning, Kukai suggests that dharanis are in fact saturated with meaning – every syllable is symbolic on multiple levels.

One of Kukai's distinctive contributions was to take this symbolic association even further by saying that there is no essential difference between syllables of mantras and sacred texts, and those of ordinary language. If one understood the workings of mantra, then any sounds could be a representative of ultimate reality. This emphasis on sounds was one of the drivers for Kukai's championing of the phonetic writing system, the kana, which was adopted in Japan around the time of Kukai. He is generally credited with the invention of the kana, but there is apparently some doubt about this story amongst scholars.

This mantra based theory of language had a powerful effect on Japanese thought and society which up until Kukai's time had been dominated by imported Chinese culture of thought, particularly in the form of the Classical Chinese language which was used in the court and amongst the literati, and Confucianism which was the dominant political ideology. In particular Kukai was able to use this new theory of language to create links between indigenous Japanese culture and Buddhism. For instance he made a link between the Buddha Mahavairocana and the Shinto sun Goddess Amaterasu.

Since the emperors were thought to be descended form Amaṭerasu, Kukai had found a powerful connection here that linked the emperors with the Buddha, and also in finding a way to integrate Shinto with Buddhism, something that had not happened with Confucianism. Buddhism then became essentially an indigenous religion in a way that Confucianism had not. And it was through language, and mantra that this connection was made. Kukai helped to elucidate what mantra is in a way that had not been done before: he addresses the fundamental questions of what a text is, how signs function, and above all, what language is. In this he covers some of the same ground as modern day Structuralists and others scholars of language, although he comes to very different conclusions.

In this system of thought all sounds are said to originate from "a" – which is the short a sound in father. For esoteric Buddhism "a" has a special function because it is associated with Shunyata or the idea that no thing exists in its own right, but is contingent upon causes and conditions. In Sanskrit "a" is a prefix which changes the meaning of a word into it's opposite, so "vidya" is understanding, and "avidya" is ignorance.

The letter a is both visualized in the Siddham script, and pronounced in rituals and meditation practices. In the Mahavairochana Sutra which is central to Shingon Buddhism it says: Thanks to the original vows of the Buddhas and Bodhisattvas, a miraculous force resides in the mantras, so that by pronouncing them one acquires merit without limits".

Mantra in Indo-Tibetan Buddhism

Conze distinguishes three periods in the Buddhist use of mantra. Initially Buddhists used mantra as protective spells to ward of malign influences. Despite a Vinaya rule which forbids monks engaging in the practice of chanting mantras for material gain, there are a number of such verses or *'parittas'* in the early Buddhist scriptures. The Atanatiya Sutta (a.tana.tiya) in of the Pali Canon (Digha Nikaya Sutta 32) is a collection of these parittas.

The verses in this sutta are the means by which people may *"dwell guarded, unharmed and at ease"*, being particularly effective against the malign influence of non-human beings. Other examples include the Ratana Sutta (Sn 222ff) and Khandha Parrita (AN 4.67). According to Pali commentary, the very well known Metta Sutta was originally taught as a protection against some disruptive tree spirits that were making life very uncomfortable for a group of ascetic monks.

However even at this early stage, there is perhaps something more than animistic magic at work. Particularly in the case of the Ratana Sutta the efficacy of the verses seems to be related to the concept of 'truth'. Each verse of the sutta ends with "by the virtue of this truth may there be happiness".

Later mantras were used to guard the spiritual life of the chanter, and sections on mantras began to be included in some Mahayana sutras such as the White Lotus Sutra, and the Lankavatara Sutra. The scope of protection also changed in this time.

In the Sutra of Golden Light the Four Great Kings promise to exercise sovereignty over the different classes of demi-gods, to protect the whole of Jambudvipa (the India sub continent), to protect monks who proclaim the sutra, and to protect kings who patronise the monks who proclaim the sutra. The apotheosis of this type of approach is the Nichiren school of Buddhism, founded in 13th century Japan, and which distilled all Buddhist practice down to the worship of the White Lotus Sutra through recitation of the daimoku: "Nam myo ho renge kyo" which translates as "Homage to the White Lotus Sutra".

Thirdly mantra began, in the 7th century, to take centre stage and become a vehicle for salvation in their own right. Tantra started to gain momentum in 6th and 7th century, with specifically Buddhist forms appearing as early as 300CE. *Mantrayana* was an early name for what is now known as *vajrayana*, which gives us a hint as to the place of mantra in Indo-Tibetan Buddhism. The aim of vajrayana practice is to give the practitioner a direct experience of Reality, of things as they really are. Mantras function as symbols of that reality, and different mantras are different aspects of that reality – for example wisdom or compassion.

Mantras are always associated with a particular deity, with one exception being the prajnaparamita mantra associated with the Heart Sutra. One of the key vajrayana strategies for bringing about a direct experience of reality is to engage the entire psycho-physical organism in the practices. In Buddhist analysis the person consists of body, speech and mind.

So a typical sadhana or meditation practice might include mudras, or symbolic hand gestures, or even full body prostrations; the recitations of mantras; as well as the visualization of celestial beings and visualizing the letters of the mantra which is being recited. Clearly here mantra is associated with speech. The meditator may visualize the letters in front of themselves, or within their body. They may pronounced out loud, or internally in the mind only.

Om Mani Padme Hum

The most famous mantra of Buddhism is Om mani padme hum, the six syllable mantra of the Bodhisattva of compassion Avalokiteshvara (Sanskrit: *Chenrezig*, Tibetan). This mantra is particularly associated with the four-armed Shadakshari form of Avalokiteshvara. Dalai Lama is said to be an incarnation of Avalokiteshvara, and so the mantra is revered by his devotees.

Donald Lopez gives a good discussion of this mantra and it's various interpretations in his book *Prisoners of Shangri-LA: Tibetan Buddhism and the West*. Lopez is an authoritative writer and challenges the stereotypical analysis of the mantra as meaning "The Jewel in the Lotus", an interpretation that is not supported by either a linguistic analysis, nor by Tibetan tradition, and is symptomatic of the Western Orientalist approach to the 'exotic' East. He suggests that Manipadme is actually the name of a bodhisattva, a form of Avalokiteshvara who has many other names in any case including Padmapani or *lotus flower in hand*.

The Brahminical insistence on absolutely correct pronunciation of Sanskrit broke down as Buddhism was exported to other countries where the inhabitants found it impossible to reproduce the sounds. So in Tibet for instance, where this mantra is on the lips of many Tibetans all their waking hours, the mantra is pronounced Om mani peme hum.

Some other mantras used by Tibetan Buddhists:

- *Om wangishwari hum* This is the mantra of the Mahabodhisattva Manjusri (Jambeyang) Buddha in his wisdom aspect.
- *Om mani padme hum* The mantra of Chenrezig, Mahabodhisattva, the Buddha in his compassion aspect.
- *Om vajrapani hum* The mantra of the Buddha as Protector of the Secret Teachings. *i.e* as the Mahabodhisattva Channa Dorje (Vajrapani).
- *Om ah hum vajra guru pema siddhi hum* The mantra of the Vajraguru Guru Padma Sambhava who established Mahayana Buddhism and Tantra in Tibet.
- *Om tare tuttare ture swaha* The mantra of Jetsun Dolma, the Mother of the Buddhas.
- *Om amarani jiwantiye swaha* The mantra of the Buddha of limitless life: the Buddha Amitayus (Tibetan Tsepagmed) in celestial form.
- *Om dhrum swaha* The purificatory mantra of the mother Namgyalma.
- *Om ami dhewa hri* The mantra of the Buddha Amitabha (Hopagmed) of the Western Buddhafield, his skin the colour of the setting sun.
- *Om ah ra pa tsa na dhih* Mantra of the "sweet-voiced one" Jambeyang or Manjusri, the Buddha in his wisdom aspect.
- *Hung vajra phat* The mantra of the Mahabodhisattva Vajrapani in his angry (Dragpo) form.

Mantra in Other Traditions or Contexts

Transcendental Meditation, known simply as 'TM', uses simple two syllable mantras as a meditative focus. TM was founded by Mahesh Yogi. According to TM, the practice can result in a number of material benefits such as relaxation, reduced stress, better health; but it can also benefit the world by reducing violence and generally improve quality of life. While the founder was a well versed Hindu, the TM tries to separate itself from that tradition these days.

Mantra practice has also been taken up by various New Age groups, although this is typically out of context, and from the point of view of a Hindu or Buddhist practitioner, lacks depth. The mere repetition of syllables can have a calming effect on the mind, but the traditionalist would argue that mantra can be an effective way of changing the level of ones consciousness when approached in traditional way.

Meditation usually refers to a state of extreme relaxation and concentration, in which the body is generally at rest and the mind quieted of surface thoughts. Several major religions include ritual meditation; however, meditation itself need not be a religious or spiritual activity. It is widely thought to be of Indian origin. The earliest mention of meditation comes from the Vedas, and Yoga, with the Sanskrit term 'Dhayana'.

Another form of meditation is more closely akin to prayer and worship, wherein the practitioner turns spiritual thoughts over in the mind and engages the brain in higher thinking processes. The goal in this case is the receipt of spiritual insights and understanding. From the point of view of psychology, meditation can induce, or is itself, an altered state of consciousness.

Strategies Common to Many Forms

Meditation generally involves avoiding wandering thoughts and fantasies, and calming and focusing the mind. Meditation does not necessarily require effort and can be experienced as "just happening". Physical postures include sitting cross-legged, standing, lying down, and walking (sometimes along designated floor patterns). Quiet is often desirable, and some people use repetitive activities such as deep breathing, humming or chanting to help induce a meditative state.

Meditation can be done with the eyes closed (as long as one does not fall asleep), or with the eyes open: focusing the eyes on a certain point of an object or image, and keeping the eyes constantly looking at that point.

Purposes of Meditation

The purposes for which people meditate vary almost as widely as practices. It may serve simply as a means of relaxation from a busy daily routine, or even as a means of gaining insight into the nature of reality or of communing with one's God. Many have found improved concentration, awareness, self-discipline and equanimity through meditation. The self-disciplining aspect of meditation plays a central role in Hinduism and Zen Buddhism.

Samadhi

In the *Samadhi* or *Shamatha*, or concentrative, techniques of meditation, the mind is kept closely focused on a particular word, image, sound, person, or idea. This form of meditation is often found in Hindu Yoga and Buddhist traditions, as well as in Medieval Christianity, Jewish Kabbalah, and in some modern metaphysical schools. Eknath Easwaran developed another, related, method. He called it "passage meditation"—silent repetition in the mind of memorized inspirational passages from the world's great religions. Easwaran believe that, "The slow, sustained concentration on these passages drives them deep into our minds; and whatever we drive deep into consciousness, that we become." It should be noted that this concept of *Japa Mantra*, or repetitive mantra, was not initially conceived of by Easwaran and is a long-standing tradition of Hinduism dating back to Vedic times.

Mindful Awareness Traditions

Vipassana and anapanasati are parts of broader notion of mindful awareness, which is part of the Noble Eightfold Path, the ultimate goal in Buddhism that leads to Enlightenment, and expounded upon the in the Satipatthana sutta. While in anapanasati meditation the attention is focused on the breath, in vipassana the mind is instead trained to be acutely aware of not only breathing, but all things that one comes to experience.

The concept of vipassana works in believing that the meditator's mind will eventually take note of every physical and mental experience "real-time" or as it happens, the goal being that it will gradually reveal to the practitioner how one's mind unknowingly attaches itself to things that are impermanent in nature. Thus, when such things cease to exist, one experiences the suffering from its loss. This very wisdom, achievable solely through vipassana meditation, in turn gradually frees one's mind from the attachment that is the root of suffering. For one practicing this form of meditation, it is also very important to note that the wisdom presents itself into the mind of the meditator only when he or she is NOT thinking but yet possessing an acute awareness of what goes on in his or her mind, body and surroundings. The nature of vipassana is sophisticated; one may desire a veteran instructor to provide initial guidance.

In other words, in Vipassana (insight, or seeing things as they are) meditation, the mind is trained to notice each perception or thought that passes without "stopping" on any one. This is a characteristic form of meditation in Buddhism, especially in some Theravada traditions, and is also a component of Zazen, the term for meditation practice in Zen. However, in at least some forms of vipassana, one does not attend to whatever perceptions arise, but purposely moves one's attention over their body part by part, checking for perceptions, being aware and equanimous with them, and moving on. This form of meditation has some resemblance with 'choiceless awareness', the kind of meditation that J. Krishnamurti addressed.

Jesus-style Meditation

Jesus, according to the New Age writing *The Urantia Book*, often left his apostles and the crowds to distance himself in the wilderness areas of Palestine to engage in long periods of dynamic-mind, spiritual meditation wherein he communicated with God. The 40 days following his baptism were spent in such a manner.

Health Applications of Meditation

Meditation has entered the mainstream of health care as a method of stress and pain reduction. For example, in an early study in 1972, transcendental meditation was shown to effect the human metabolism by lowering the biochemical byproducts of stress, such as lactate (lactic acid), and by decreasing heart rate and blood pressure and inducing favourable brain waves.

As a method of stress reduction, meditation is often used in hospitals in cases of chronic or terminal illness to reduce complications associated with increased stress including a depressed immune system. There is a growing consensus in the medical community that mental factors such as stress significantly contribute to a lack of physical health, and there is a growing movement in mainstream science to fund and do research in this area. Dr. James Austin, a neurophysiologist at the University of Colorado, reported that Zazen or Zen meditation rewires the circuitry of the brain in his landmark book *Zen and the Brain*.

This has been confirmed using sophisticated imaging techniques which examine the electrical activity of the brain. Dr. Herbert Benson of the Mind-Body Medical Institute, which is affiliated with Harvard and several Boston hospitals, reports that meditation induces a host of biochemical and physical changes in the body collectively referred to as the "relaxation response." The relaxation response includes changes in metabolism, heart rate, respiration, blood pressure, and brain chemistry.

The Urantia Book indicates that relaxation is similarly a result of dynamic-mind meditation, thus its health benefits should be the same. However this has not been confirmed by scientific investigations.

Specific traditions-specific classifications include:

- Observation (*e.g.*, exploring the mind and all its thoughts)

- Focus (*e.g.*, exploring one thought to the exclusion of all else)
- Trance (experiencing emptiness)
- Theravada Buddhist practice involves both Samadhi and Vipassana, as well as the developing of "loving kindness" (Metta).
- Zen Buddhism practices Zazen, similar to Vipassana.
- Most Abrahamic traditions practice forms of meditation that use their God, Saints and/or Prophets as concentration focus
- Some people, including the controversial Guru Rajneesh (also known as "Osho"), taught forms of "Dynamic Meditation" that involve violent exercise and hyperventilation, akin to aerobic exercise or those like the Sufi whirling.
- Meditation based on questioning "Who am I" draws from various traditions, especially Vipassana, Insight Meditation, Zazen, and with the express purpose of getting to know one's true nature, and/or experiencing Kensho, Satori, Enlightenment. This method teaches to be wary of trance-like states of relaxation, and advocates intense inquiry into the nature of thought, mind, ego, self, and desire.
- Jesus-style, dynamic-mind, meditation engages the creative consciousness and full awareness. The mind and thinking processes are expanded rather than shut down or dulled. The only requirements are sincerity, persistence, and God-consciousness. *The Urantia Book* cautions against the use of "physical fatigue, fasting, psychic dissociation, profound aesthetic experiences, vivid sex impulses, anxiety, rage, and wild dancing" to initiate "mystic communion". It states that "the trancelike state of visionary consciousness should not be cultivated as a religious experience."

Tantra (Sanskrit: loom), Tantric Yoga or Tantrism is any of several esoteric traditions rooted in Hindu and Buddhist philosophy. Extolled as a short-cut to self-realization and spiritual enlightenment by some, tantric rites are often rejected as dangerous by most orthodox Hindus. There are two "paths" in Tantra: *dakshinachara* (also known as *samayachara*), the "Right-Hand Path", and *vamachara*, the "Left-Hand Path". The latter is associated with many ritual practices that go against the grain of mainstream Hinduism, including sexual rituals, consumption of alcohol and other intoxicants, animal sacrifice and flesh-eating. The terms *Left-Hand Path* and *Right-Hand Path* have been adopted by Western occultists.

According to another popular view, Tantra is classified as either red (rajas or heat, fire, restlessness, anger), black (tamas or darkness, ignorance, stagnation) or white (sattwa pure, moderate, divine). These correspond to three Hindu conceptions of the qualities of existence first posited in Samkhya. Some Tantric aspirants simply feel the union is accomplished internally and with spiritual entities of various kinds. For this reason, almost all Tantrik writing has a gross, higher and subtle meaning. This tripartite system of understanding readily obscures the true purport of many passages for those without the necessary background or deeper understandings so crucial to Tantra. Thus, a 'union' could mean the actual act of sexual intercourse, ritual uniting of concepts through chanting and sacrifice, or realization of one's true self in the cosmic joining of the divine principles of Shiva and Shakti in Para Shiva.

The Hindu Tantra Tradition

According to John Woodroffe, one of the foremost Western scholars on Tantra, and translator of its greatest works (including the Mahanirvana Tantra):

> *"The Indian Tantras, which are numerous, constitute the Scripture (Shastra) of the Kaliyuga, and as such are the voluminous source of present and practical orthodox "Hinduism." The Tantra Shastra is, in fact, and whatever be its historical origin, a development of the Vaidika Karmakanda, promulgated to meet the needs of that age. Shiva says: "For the benefit of men of the Kali age, men bereft of energy and dependent for existence on the food they eat, the Kaula doctrine, O auspicious*

one! is given" (Chap. IX., verse 12). To the Tantra we must therefore look if we would understand aright both ritual, yoga, and sadhana of all kinds, as also the general principles of which these practices are but the objective expression."

Introduction to Sir John Woodroffe's translation of "Mahanirvana Tantra."

The word "tantra" means "treatise", and is applied to a variety of mystical, occult, medical and scientific works as well as to those which we would now regard as "tantric". Most tantras were written between the 10th and 14th centuries CE.

While Hinduism is typically viewed as being Vedic, the Tantras are not considered part of the orthodox Hindu/Vedic scriptures. They are said to run alongside each other, The Vedas of orthodox Hinduism on one side and the Agamas of Tantra on the other. However, the practices, mantras and ideas of the Atharva Veda are markedly different from those of the prior three and show signs of powerful non-Aryan influence. Indeed, the Atharva Veda is cited by many Tantra texts as a source of great knowledge. It is notable that throughout the Tantras, such as the Mahanirvana Tantra, they align themselves as being natural progressions of the Vedas. Tantra exists for spiritual seekers in the age of Kaliyuga, when Vedic practices no longer apply to the current state of morality and Tantra is the most direct means to realization.

Thus, aside from Vajrayana Buddhism, much of Tantric thought is Hindu Tantra, most notably those that council worship of Lord Shiva and the Divine Mother, Kali. A Tantra typically takes the form of a dialogue between the Hindu gods Shiva and Shakti/Parvati, being that Shiva is known in Hinduism as being 'Yogiraj' or 'Yogeshwara,' 'The King of Yoga' or 'God of Yoga' and that his consort is known to be his perfect feminine equal. Each explains to the other a particular group of techniques or philosophies for attaining *moksha* (liberation/ enlightenment), or for attaining a certain practical result. [Agamas are Shiva to Shakti, and Nigamas are Shakti to Shiva.]

This extract from the beginning of the *Yoni Tantra* (translated by Mike Magee) gives an idea of the style.

Seated upon the peak of Mount Kailasa the God of Gods, the Guru of all creation was questioned by Durga-of-the-smiling-face, Naganandini.

"Sixty-four tantras have been created O Lord, tell me, O Ocean of Compassion, about the chief of these."

Mahadeva said:

"Listen, Parvati, to this highly secret one, Dearest. Ten million times have you wanted to hear this. Beauteous One, it is from your feminine nature that you continually ask me. You should conceal this by every effort. Parvati, there is mantra-pitha, yantra-pitha and yoni-pitha. Of these, the chief is certainly the yoni-pitha, revealed to you from affection."

History of Tantra

Tantra as a post-Vedic Hindu Yogic movement began in North India and flourished in the middle ages before declining in the nineteenth century, partly as a result of persecution by the British and orthodox Hindus, and partly, perhaps, because of the increasing popularity of Bhakti yoga amongst the masses.

Legend ascribes the origin of Tantra to Dattatreya, a semi-mythological yogi and the assumed author of the *Jivanmukta Gita* ("Song of the liberated soul"). Others see Lord Adinath, or Shiva, as the first Guru of Tantra. Things become a little more clear with Matsyendranath ("Master of fish"-so-called either because he was a fisherman, or, less probably, because he discovered a tantra inside a fish). He is accredited with authorship of the *Kaulajnana-nirnaya*, a voluminous ninth-century tantra dealing with a host of mystical and magical subjects, and occupies an important position in the Hindu tantric lineage, as well as in Tibetan Vajrayana Buddhism. His disciple, Gorakhnath, founded laya yoga. Hatha Yoga was penned by Swami Swatamarama as the secrets of Lord Adinath (another name for Shiva) in the 15th century. Tantra evolved into a number of orders (*sampradaya*) and diverged into so-called "left-hand tantra" (*varma marg*), in which sexual yoga and other antinomian practices occurred, and "right-hand tantra", in which such practices were merely visualized. Both groups, but in particular the left-hand tantriks, opposed many features of orthodox Hindu culture, particularly the caste system and patriarchy. Despite this, Tantra was accepted by some high-caste Hindus, most notably the Rajput princes. Hindu tantra even briefly enabled a yogic/sufi synthesis among some Indian Muslims. Nowadays Tantra has a large, though not always well-informed, following worldwide.

Buddhist and Hindu Tantra, though having many similarities from the outside, do have some clear distinctions. Scholars are unable to determine whether the Hindu or the Buddhist version of Tantra appeared first in history. Buddhist Tantra is always part of the Mahayana school of Buddhism, which has as main aim to help all sentient beings becoming free from problems (Dukkha), in order to achieve this aim, one should try to achieve Buddhahood oneself, in order to be the most profound teacher for others. Buddhist Tantra spread out from (North) India, chiefly to Tibet, where it became known as the Vajrayana school of Buddhism. It also had some influence on Chinese and Japanese Buddhism (notably Shingon)

Tantric Practices

Because of the wide range of groups covered by the term "tantra", it is hard to describe tantric practices definitively. The basic practice, the Hindu image-worship known as "puja" may include any of the elements below.

Mantra and Yantra

As in all of Hindu and Buddhist yogas, mantras plays an important part in Tantra, not only for focusing the mind, often through the conduit of specific Hindu gods like Shiva, Ma Kali (mother Kali, another form of Shakti) and even Ganesha, the elephant-headed god of wisdom. Similarly, puja will often involve concentrating on a yantra or mandala.

Identification with Deities

Tantra, being a development of Atharva Vedic and pre-Brahmanical thought, embraced the Hindu gods and goddesses, especially Shiva and Shakti, along the Advaita (nondualist Vedic) philosophy that each represents an aspect of the ultimate

Param Shiva, or Brahman. These deities may be worshipped externally (with flowers, incense etc.) but, more importantly, are used as objects of meditation, where the practitioner imagines him-or herself to be the deity in question. The ancient devadasi tradition of sacred temple-dance, seen in the contemporary Bharata Natyam is the example of such meditation in movement. The divine love is expressed in Sringara and Bhakti.

Concentration on the Body

Tantriks generally see the body as a microcosm; thus in the *Kaulajnana-nirnaya*, for example, the practitioner meditates on the head as the moon, the heart as the sun and the genitals as fire. Many groups hold that the body contains a series of energy centres (chakra-"wheel"), which may be associated with elements, planets or occult powers (siddhi). The phenomenon of Kundalini, a flow of energy through the chakras, is controversial; most writers see it as essential to Tantra, while others regard it as unimportant or as an abreaction. As it is, kundalini is nothing but the flow of the central sushumna nadi, a spiritual current, that, when moving, opens chakras, and is fundamental to the siddhi concept that forms a part of all tantra, including Hatha yoga.

Sexual intercourse is not at all a part of all tantric practice, but it is the definitive feature of left-hand Tantra. All tantra states that there were certain groups of personalities who were not fit for certain practices. Tantra was personality specific and insisted that those with pashu-bhava (animal disposition), which are people of dishonest, promiscuous, greedy or violent natures who ate meat and indulged in intoxication, would only incur bad Karma by following Tantrik paths without the aid of a Guru who could instruct them on the correct path. There are three types of Tantric Sex: White, Gray, and Black. White Tantra never ejaculates nor reaches orgasm in order to awaken consciousness. Gray Tantra elongates the sexual act, and sometimes concludes with orgasm/ejaculation, but without any longing towards awakening consciousness. Black Tantra always concludes with orgasm/ejaculation in order to awaken consciousness. It is said that White Tantra awakens consciousness to the absence of desire, while Black Tantra awakens consciousness into desire. In Buddhist tantra, actual ejaculation is very much a taboo, as the main goal of the sexual practice is to use the sexual energy towards achieving full enlightenment, rather than ordinary pleasure.

Sexual intercourse, preferably with a low-caste partner, was one method by which traditional left-hand practice forced practitioners to confront their conditioned responses. Others include the eating of meat (particularly beef and pork) and drinking of alcohol. Fear has also been used as a method to break down conditioning; rites would often take place in a cremation ground amidst decomposing corpses. This, of course, also falls under the prerequisite of the practitioner's nature, in such cases demanding a vir-(heroic) or even devya-(godlike)-bhava (disposition of purity, self-control, suppression of pride, respect to parents and guru and often celibacy).

Tantra in the Modern World

Tantra is used in the West, as a general term which relates to sexual practice as a spiritual evolutionary scheme. There are in fact many different approaches as to how this manifests in American society. There have been many civilizations which have deified sexuality as the most approximate expression of cosmic love or God. Regardless, the point is that tantra is moldable. It changes with each moment and environment. It especially depends on the nature of the practitioner.

In traditional pockets of Tantric practice in India, such as in Assam near the venerated Hindu temple of Kali, Kammakha, in parts of West Bengal, in Siddhanta temples of South India, and in Kashmiri Shiva temples up north, Tantra has retained its true form. Its variance in practice is seen, where many tantriks are known to frequent cremation grounds in attempts to transcend their worldly attachment to life, and others are assuredly performing still more arcane acts. But what is common to them all is the intense secrecy in which their secrets are kept and the almost godlike reverence paid to the Guru, who is seen as a the pinnacle of Tantra.

It would be safe to say that every single Hindu Tantra Yogin in India is a Shiva and/or Shakti worshipper, and the more wide-spread practices to which all Hindus commit themselves, like pooja and worship through devotion, are maintained while more occult yogic practices involving sacred rites continue.

Tibet too has a very strong Buddhist Tantric background which continues, albeit many have been transplanted to monasteries in India, but can be said to widely cleave to the right-hand path, in contrast to the more varied Hindu counterparts. Modern Tantra may be divided into practices based on Hinduism and Buddhism, Indian and Tibetan, traditions. In America, a mutilated and extremely narrow-minded, sensationalist approach encompassing only a misguided thinking about "sacred sexuality," with little reference to its true practice, has captured the Western mind. Real Tantra involves much more than mere wizardry or sexual titillation: like the rest of Yoga (Hindu and Buddhist), it requires self-analysis and conquering of material ignorance, often through the body, but always through a pure outlook of the mind.

'Real Tantra' is about transforming one's sexual energy into spiritual progress, and has nothing to do with 'sex just for fun'. Those without a guru or lacking in discipline of the mind and body are unfit. It is telling that a Tantrik in West Bengal, a devotee of the Hindu goddess Kali, once said that '*those most fit for Tantra almost never take it up, and those least fit pursue it with zeal.*' Mantra is a syllable or string of syllables, from the Sanskrit language, first found in Hinduism. Mantras may or may not conform to grammatical rules. Their use varies with the school and philosophy associated with the mantra.

They are primarily used as spiritual conduits, words and vibrations that instill one-pointedness in the devotee. Other purposes have included religious ceremonies to accumulating wealth, avoiding danger, or eliminating enemies. Mantras originated in India with Vedic Hinduism and were later adopted by Buddhists and Jains, now popular in various modern forms of spiritual practice which are based on practices of these Indian religions.

7

Path to Self-realization in Hindu Life

In the preceding two chapters, we made a study of the state of self-realization and the state of ignorance respectively. Our analysis of the state of self-realization, in the first chapter, helped us to understand that the ultimate reality behind this universe is *Brahman*, the pure consciousness in *jiva* is *Atman* and that they are fundamentally one. Besides, it is a non-dual state of unbroken bliss, in which the seeker of *Brahman* experiences the absolute in his self and views the whole world form the oneness-perspective of *Brahman*.

Our study of the state of ignorance, in the second chapter, brought to light the nature of phenomenal existence, in which a person is caught up with the *Ishvara—jiva* experience, which is brought about by the superimposing activity of the cosmic and individual *maya* on *Brahman*. As a result, the non-dual *Brahman* identical with *Atman* is experienced by the *jiva* in three cosmic states, *viz.*, the *Virat*, the *Hirayanagarbha*, and *Ishvara*, and three individual states, *i.e.*, the *Vishva*, the *Taijasa* and the *Pranja*. The second chapter also pointed out to the fact that only way to experience self-realization is to transcend the state of ignorance in all its aspects.

The removal of ignorance is the condition on which the identity experience between *Brahman* and *Atman* can happen. Thus, the negation of ignorance is a *sine qua non* for *Brahman*-experience. Shankara says: "The following knowledge [the knowledge of *Brahman*, *i.e.*, self-realization] does not arise without negating the previous one [superimposed knowledge that is characteristic of the state of ignorance] [as] the knowledge of the rope does not come without destroying that of the snake in the snake-rope." Thus, for Shankara, man's movement towards authentic self-realization is an ascending movement, like climbing a ladder, in which the lower stages are given up, when the higher stage is attained. The path that leads to self-realization is called *Brahmajijnasa*, which is the topic of our consideration in the first section of this chapter. The second section deals with the *jnana* path to self-realization proposed by Shankara. In the third section we elaborate the nature of release (*Samadhi*) and the nature of the self-realized man (*Brahmajnani*).

Nature and Methods of the Pathi

In this section we clarify the nature of *Brahmajijnasa*. Besides, we will also look into the methods proposed by Shankara in the attainment of the state of self-realization.

Nature of Brahmajijnasa

Here, we analyze the nature of *Brahmajijnasa*, the path to self-realization, by highlighting its meaning. Having spelt out the meaning of *Brahmajijnasa*, we would explicate its nature by distinguishing it from *Brahmanubhava*, the self-realization.

Meaning of Brahmajijnasa: *Brahmanubhava* is an immediate and direct experience. So there is no need of any means to attain it. Besides, it is an absolute and trans-empirical experience, that it is not possible to attain it by the use of any empirical or relative means. But, though we cannot bring about *Brahmanubhava*, still we can remove ignorance and pave the way for the realization of the self. The non-realization of the self is as a result of ignorance, and the removal of ignorance is all that we need to do in order that *Brahmanubhava* can happen in us.

"The attainment of the highest [*Brahmanubhava*]", says Shankara, "means merely the removal of ignorance and nothing more". In darkness, though the objects are in contact with the senses, still we are not able to perceive them because of the presence of darkness. When we remove the darkness, we are able to perceive the objects as they are. For example, in semi-darkness a stump of tree appears like a man; when light comes in, we can recognize the real nature of object in question, *i.e.*, the tree stump. This is much the same in *Brahmanubhava*. It is due to ignorance that the absolute reality, the *Brahman*, is seen as the world of phenomena. All that is required to attain self-realization is to remove ignorance that blinds one to see one's own nature. It is the wrong knowledge of the self that obstructs the perfect knowledge of the self, just as the perception of the silver in the pearl-shell obstructs the perception of the pearl-shell. When the obstruction of wrong knowledge is removed, the real nature of the self becomes self-revealed.

Thus, for Shankara, "except for the removal of ignorance of the superimposition of the non-self on the self, no effort, whether bodily or mental is necessary for the realization of the self. *Brahmajijnasa* is the process of attaining *Brahmanubhava*. The aspirant undertakes this process in order to remove ignorance that prevents him from attaining self-realization. The term '*Brahmajijnasa*' derives from two Sanskrit terms, *viz.*, '*Brahman*' which means 'absolute reality' and '*jijnasa*' which means 'wish to know'. So literally *Brahmajijnasa* means 'a wish to know *Brahman*, the absolute reality'. A wish emerges spontaneously from the knowledge that something is achievable by effort and that when achieved it will lead to desired results. Thus, a wish implies all the efforts involved in achieving the desired result.

Thus, by implication the term '*Brahmajijnasa*' means the deliberation on the upanishadic texts about *Brahman*, undertaken by the aspirant for getting direct knowledge of *Brahman*. 'Wish' by implication means the deliberation resulting from the wish to know *Brahman* and 'knowledge' means the special kind of direct knowledge of *Brahman*. Both the verbs 'to wish' and 'to know' are transitive and must have their objects: 'wish' has knowledge for its object and 'knowledge' has *Brahman* as its object. But man must know first in order that he can wish the knowledge of *Brahman*. This knowledge is provided by the scriptures, *viz.*, indirect, mediate and apparent knowledge that one gains from the study of scriptures. But the knowledge culminating in the direct realization of *Brahman* is the object of 'the wish'. This knowledge of *Brahman* is as the result of the wish to know *Brahman* (*Brahmajijnassa*). It involves intellectual deliberation on *Brahman*, moral life that opens one to the experience *Brahman* and the deep meditation on the mediate and indirect knowledge arrived at by the study of the scriptures. Thus, the knowledge arrived at in *Brahmajijnasa* is a mature knowledge, which culminates in the direct realization of *Brahman*.

From what we have said in relation to the Sanskrit origins of the word '*Brahmajijnasa*', it is clear that this term stands for all the efforts an aspirant makes to move from the state of ignorance to the state of self-realization. It would involve all the different means used by the aspirant to attain self-realization. It would include the physical and moral efforts an aspirant puts into purify his body and mind. It consists firstly, the disciplining of learning and deliberation one acquires in the intellectual level, to study and understand the import of the scriptural statement about *Brahman*. Secondly, the guidance one gets from his *Guru*, in clarifying the doubts that may arise in the process of scriptural study. Thirdly, the meditation the aspirant undertakes to practice in order to have direct experience of *Brahman*. Thus, *Brahmajijnasa* is the path to self-realization. It would embrace the different means used and efforts made by the aspirant to transcendent the duality of the empirical order and attain the identity consciousness. Now that we have clarified the meaning of *Brahmajijnasa*, we can make an attempt to differentiate it from *Brahmanubhava*, the goal of *Brahmajijnasa*.

Brahmajijnasa and Brahmanubhava

Brahmajijnasa is a process, in which the seeker in an ascending manner moves towards *Brahman* by removing the state of ignorance that is characteristic of the phenomenal existence. The aspirant having taken upon himself the process *Brahmajijnasa* goes through its different stages. These help him to get rid of ignorance, which is the cause of duality and multiplicity. When ignorance is removed, and with it all multiplicity, the truth about oneself dawns. For Shankara, the *jnana* path is the fundamental aspect of *Brahmajijnasa*, as only knowledge is able to remove ignorance. In *jnana* path, by hearing is removed the doubt from an unprepared mind that the Upanishads cannot impart the knowledge of *Brahman*. Reflection removes the doubt that the self and *Brahman* cannot be one, especially by giving logical and reasonable arguments. Through meditation the mind is withdrawn from all distractions and things other than *Brahman*.

This process of *Brahmajijnasa* gradually removes all traces of ignorance from the aspirant and thereby opens the way to *Brahmanubhava*, the ultimate liberation. Hearing, reflection and meditation generate knowledge by the removal of ignorance. When all consequences of ignorance, such as duality and differences are removed, there dawns the self-knowledge, which is self-realization. *Brahmajijnasa* does not cause the absorption into *Brahman*, because it deals only with what is known from hearing. The knowledge attained through *Brahmajijnasa* is an indirect knowledge. The knowledge attained through the process *Brahmajijnasa*, by concentrating on the import of the scriptures is the highest intellectual knowledge possible. Nevertheless, *Brahmanubhava* is not attained with *Brahmajijnasa*, but the former is above and beyond the latter, since the direct and immediate experience of one's self. Dr. Smet remarks the following about *Brahmajijnasa* and its relationship to *Brahmanubhava*:

> *Brahmajijnasa is only a preparation, a progressive removal of obstacles to knowledge, the protracted suicide of ajnana (ignorance). Vidhya (true knowledge) cannot be a result, but a direct, independent realization, and awakening or reality, an intuition suddenly dawning upon the seeker. It is direct knowledge, while the result affected by the whole Brahmajijnasa can be nothing higher than an indirect knowledge. The occurrence of Brahmanubhava is simultaneous with the removal of ignorance from the seeker. When, ignorance is removed, by the process of Brahmajijnasa, there follows*

immediately the truth about oneself, i.e., one's absolute absorption into Brahman. At this stage the indirect knowledge of reflection (vichara) gives way to right discrimination (viveeka). Now, there is no activity and the intellect of the aspirant sees through the appearance, recognizing his oneness with Brahman. When this happens, the aspirant knows that he is Brahman and all duality and multiplicity disappears, since he realizes the fact that all is Brahman. This realization of one's absorption into and identity with Brahman is Brahmanubhava, the true liberation. The Brahmajijnasa is a preparation, while Brahmanubhava is the end.

Now that we have clarified the nature of *Brahmajijnasa*, by exploring its meaning and distinguishing it from self-realization, we could move on to speak of the methods used by the Advaitin, in the process of *Brahmajijnasa*.

Methods of Brahmajijnasa: Shankara's Advaita Vedanta recognizes a dualism in the process of *Brahmajijnasa*. In other words, he speaks of an indirect and a direct method in *Brahmajijnasa*. In this section, we could clarify each of these methods and the reason why Shankara advocated these dual methods. Besides, we would also indicate the primacy of the direct method in the attainment of self-realization.

Indirect Method: By indirect methods Shankara means the path of action (*karmayoga*) and the path of devotion (*bhaktiyoga*). These open in the individual aspirant a willingness to surrender his energy in a life of service and to give with a loving heart. The willing surrender of one's life in generous service calls one to live a life of sacrifice and action. For such a life of action, it is necessary to work towards eliminating selfish motives and intentions that stem from animal instincts in man.

Besides, the aspirant must learn to do everything he does without any self-interest (*nishkamakarma*) Action, thus, done in love and faith would open the heart of the aspirant for higher level of existence. The life of sacrifice is a stream through which the divine flow can envelop the heart of the aspirant to a deeper and fuller life of love and devotion. Life of devotion prepares the seeker for deeper levels of sweetness and bliss. These transformations take place gradually, leading the seeker to still higher stages. We could consider these two indirect methods in some detail.

Karmayoga

The path of action is proposed for persons of mixed temperaments, *viz.*, persons who are both emotional and rational. These are energetic personalities who have an inherent orientation towards activity, work and service. *Karma* path calls for the simultaneous use of one's head and heart. A *karmayogi* dedicates his activities in love and devotion to the Lord. In the process, he sheds his personality off worldly thoughts and pleasures, gets his mind purified and his attention is turned to the contemplation of the supreme self, which is the source of his activity of love and service.

In order to attain this goal of perfect action, which will lead to the contemplation of the self, the seeker of *karma* path must let go of his personal likes (*ragga*) and dislikes (*dwesha*) in performing his action. One should not do an action because one likes it; nor should one give up certain actions because he dislikes it. One must learn to overcome one's whims and fancies in performing actions. Thus, the greatest enemies of man are his likes and dislikes, for they make even the highest type of activity meaningless. Avoidance of negative activity does not consist in doing what one does not like, but rather in performing actions without any personal preferences.

It consists in obtaining a sense of objectivity and using one's discriminating intellect to judge and decide on a proper course of action, irrespective of one's likes and dislikes. In this manner, one should be able to remove negative activity from one's life. Once a person brings about a sense of objectivity and steadiness by removing acts by likes and dislikes, he must cultivate positive activity. In the words of Gita, positive activity consists in acting dynamically surrendering all one's actions to the Lord, allowing one's thoughts to rest on the supreme self. Besides, such action must be free from egoism, hope and feverish excitement. This verse from Gita speaks of twofold characteristics of positive activity, *viz.*, acting dynamically so as to use one's body, mind and intellect purposefully resting on the supreme Lord; and not to allow the energies produced by dynamic action be dissipated in an unproductive manner. The first characteristic involves engaging in constant action, using the body, because action builds up energy in the body, while inaction weakens a person's body. In performing the action one's thoughts must rest on the Lord, always becoming aware that the inner self is the foundation of one's actions. In this manner one builds up mental stamina and intellectual ability for work. Thus, positive activity consists in building up one's body, mind and intellect that the person is ready for genuine action. Secondly, once the energy of a person is built up by dynamic action that is resting on the Lord, the person must see that the energies thus built up should not be lost.

There are three outlets that dissipate energy, *viz.*, egoism, hope and excitement. Egoism and self-interest can vitiate a person's sense of purpose. So one must see that self-interest does not lead a person astray from the path of disinterested action (*nishkama karma*). Again hope of enjoying the fruits of one's actions can de-rout a person from good intentions. So a person seeks the path of *karma* must detest craving for enjoyment. Besides, excitement brings unrest in a person's sphere of activity. An excited person cannot be fully responsible for his duties and responsibilities.

Thus, egoism vitiates the past, by focusing on a person's past activities; the hope of enjoyment dissipates a person acting purposefully in the future; and excitement disturbs the present activity. When a seeker of the *karma* path performs dynamic action resting his thoughts on the Lord, and prevents egoism, hope and excitement to vitiate the energy in the threefold temporal dimensions of his life, he becomes a *karmayogi*. Having analyzed the path of action we could proceed to consider the path of devotion, the other indirect method.

Bhaktiyoga

Bhaktiyoga is meant for a person who is more emotional in nature, whose heart dominates his head. Such a person is more emotional, devotional and impulsive, while he is less rational, intellectual and discriminative. So persons of this nature could take recourse to *bhaktiyoga*. He often makes use of a personal God (*Ishtadeevata*) to pour out his love and devotion. He makes use of *mantras*, *bhajans* and chants to express his personal love for the Lord. His love and devotion for the Lord makes him single-minded in his attachment to the Lord. This, in turn, frees him from worldly attachments, which prepares him for deeper meditation and realization of self. When a person practices devotion to the Lord he experiences devotion in two ascending stages.

Thus we can speak of two types of devotion, *viz.*, *bheda-upasana* and *abheda-upasana*. In *bheda-upasana* a difference-in-identity is kept in view. The seeker, in this state, while enjoys the delights of the union in love, keeps himself separate. *Abheda-upasana* is a state, in which, the identity with the absolute is kept in view. Here, the aspirant gradually loses himself in the object of his love and worship. Thus, in the latter state, there is the possibility of identity consciousness, as life in the world does not seem anything more than the mental life of the seeker. Such an identity consciousness can take place in two different processes, *viz.*, the process of sinking and the process of expansion. In the sinking process, the seeker feels that his soul is placed in the all-pervasive consciousness. Feeling the immanence of the bliss, the seeker surrenders his self completely.

The complete surrender brings about psychological and spiritual transformations in the seeker. He experiences every mental modification in a new vision and meaning, as everything is viewed in relationship to the infinite. There is complete delight and deep satisfaction in the soul. The intensity of devotion in love and the inner delight and satisfaction make the finite self-consciousness to forget itself and be absorbed in the immanent infinite consciousness. The process of expansion consists in the gradual realization, in the seeker, that he is the immanent principle of the cosmos. When this realization takes over the aspirant, he does not any more feel that he is placed in the vastness of the cosmos, but rather finds the entire universe is the reflection of his own being. In other words, he feels within himself the totality of existence. In the process, his sense of finite personality dissolves into an impersonal expansive conscious existence. Having looked into the indirect methods of *Brahmajijnasa*, *viz.*, *karmayoga* and *bhaktiyoda*, we could make an attempt to study the direct method in the next section.

Direct Method: Direct method refers to the *jnana* path to self-realization. Direct method of *jnana* path is meant for a person, who is rational and whose head dominates his heart. As a result, such a person is more intellectual and discriminative, while less emotional and impulsive. He is not moved by his feelings, as his intellect is able to view events impartially. A man of intellect is not satisfied in postulating a God and worshipping Him. But a man of wisdom looks for arguments to justify such postulation.

The significant characteristic of a person with an intellect bend of mind is his ability to discriminate between the phenomenal and the transcendental. He has the propensity for deeper meditation and experience of the transcendental self. *Jnanayoga* is designed for persons of intelligence, who can ask fundamental questions about this universe and draw answers out of their reflection. The knowledge they acquire is not merely intellectual, but experiential and existential. The *jnana* path is described as '*nitya anitya viveeka vichara*', 'the discrimination between the eternal and the temporal by way of reflection'. Thus, *jnanayoga* constantly attempts to distinguish the permanent from the impermanent, the noumena and the phenomena, the transcendental and the terrestrial, and the real and the unreal, so that the seeker of knowledge can realize the ultimate knowledge of *Brahman* as identical with *Atman*. For Advaitins *jnana* path is the only way for the direct experience of *Brahman*, as it is only true knowledge that can remove ignorance.

Thus, the direct method of *jnana* path involves a deep understanding of the illusory nature of the phenomenal world and the fundamental oneness of everything in *Brahman*. Besides, it implies a discriminative consciousness that would enable the aspirant to break through the appearance and apprehend the underlying absolute reality behind the manifold world of everyday experience.

Need for and Limitation of the Indirect Method

Though Shankara was convinced of the primacy of the direct method of knowledge for the removal of ignorance, he did give a place for the indirect methods, *viz.*, the path of action and the path of devotion. He accepted them because their practical necessity, as vast majority of people in the world would not be able to give themselves directly to the demands of a life that involves reflection and intellectual activity. In other words, direct method would not be possible for everyone, as all may not be able to attain the vision of pure reason and discriminative consciousness.

But if persons have given themselves to lower levels of existence, *i.e.*, service and love, they would gradually come to desire wisdom, transcending empirical state of existence. It was Shankara's belief that, for most people, the path of knowledge can begin to have any meaning, when they have satisfied the will and the heart, as it will lead to the realization of the absolute behind the illusory and the relative. The focus of *niskama karma* on the service of others and *bhakti* on love of the supreme Lord, would make the life of real renunciation easy while living in the phenomenal world. This is probably the reason that Shankara retained the *jiva—Ishvara* ideal in his system of thought, even though he was convinced of the truth of *Brahman—Atmana* identity. Thus, Shankara proposed the indirect methods to persons dominated by emotion and action, that having satisfied the needs in the levels of emotion and action, they can move towards the higher intellectual and discriminative consciousness.

In other words, Shankara admitted the possibility of a life of love in service and service in love, only as a help to those seekers of *Brahman*, who otherwise would not have had the opportunity to open themselves to a life of reflection (*vichara*) and discrimination (*viveeka*). In speaking of the various means to remove ignorance Shankara argues that neither action nor devotion are able to remove ignorance. These two means work in the realm of ignorance. The former concentrates on performing various actions in the world of phenomena, while the latter helps the aspirant to concentrate on the supreme Lord. Both action and devotion presuppose duality and are not opposed to ignorance. A man who performs good actions and a devoted man can still be in the illusion of ignorance and may never know the true nature of *Brahman*. Therefore, action and devotion do not succeed in removing ignorance. Shankara very strongly argues for the primacy of knowledge, in the removing ignorance. Knowledge is directly opposed to ignorance. Knowledge and ignorance about a particular reality cannot coexist in a person, for knowledge drives away ignorance.

Though knowledge cannot bring about self-realization, still it can indirectly remove ignorance by its very presence. "Whether ignorance means, doubt or false knowledge", says Shankara, "it is always removable by knowledge only, but not by action in any form, for there is no contradiction between ignorance and action. Thus, indirect paths of service and love are not by themselves capable of removing ignorance, even though they can open the individual aspirant to a life of wisdom gradually. From what we have said so far, it is clear that while Shankara holding for the primacy of *jnana*, recognizes the significance of *karma* and *bhakti*, for pragmatic reasons.

In the foregoing section, we clarified the nature of *Brahmajijnasa*, the path to self-realization. We also highlighted the indirect methods of *karmayoga* and *bhaktiyoga* and their need in the process of *Brahmajijnasa* and their limitation in removing ignorance. Besides, we introduced the direct method of the *jnana* path. In the following section, we attempt to elaborate Shankara's *jnana* path and its stages in the process of *Brahmajijnasa*.

Stages of the Jnana Path

In order to study the scriptures and thereby remove ignorance, Shankara proposes certain physical, moral and intellectual preparations. These form the three stages of *Brahmajijnasa*. The physical preparation aims at helping the seeker to attain full control over his body and this done by what Vedantins call *Hathayoga*. The moral preparation has for its goal the purification of the mind, by removing all inclinations to evil. The intellectual preparation intends to grasp the full import of the scriptures with the intellectual study of the scriptural texts. We could, now, elaborate these three in detail.

Hathayoga—Physical Preparation

In this section we would like to clarify the meaning of *Hathayoga* and its main limps, *viz.*, posture (*asana*) and breath-control (*pranayama*).

Meaning of Hathayoga: The high intellectual penetration involved in the study of the scriptures implies that the seeker of *Brahmabubhava* has full control over his body. The stability of the gross body is required for its normal functioning. When it comes to preparing oneself for higher intellectual and spiritual training, there is the need to discipline the body sufficiently, so as to make it a fit instrument for the realization of *Brahman*. The system of training that prepares the body for such higher state of existence is called *Hathayoga*.

The term '*Hathayoga*' comes from two Sanskrit terms '*hatha*' and '*yoga*'. The term '*hatha*' means violence, force and oppression. The word '*yoga*' means a technique prescribed for the removal of the tendencies (*vasanas*) of the body and the mind. Thus, *Hathayoga* means physical disciplines, which an individual undertakes, which involves a certain amount of violence, force or oppression to the body. Therefore, the practice of *Hathayoga* involves a certain amount of compulsion, either administered by others or by oneself.

The aim of *Hathayoga* is to purify the tendencies of the body and mind that the intellect can begin to reason, which would enable the aspirant to study the scripture and thereby start the process of *Brahmajijnasa*. Many practices, such as different forms of self-torture, standing on one leg, holding up arms, inhaling smoke with head inverted, piercing different parts of the body with sharp instruments and similar practices are included in the *Hathayoga*.

It increases vitality in the body, gives good health and preserves great amount of energy within the aspirant, as

Hathayoga opens the aspirant for the life-process of the cosmic *prana*.

Limps of Hathayoga

Though there are many practices included in *Hathayoga*, its two main limps are posture (*asana*) and breath-control (*pranayama*), to which we turn our attention in this section.

Posture: The first limp of *Hathayoga* is posture (*asana*). It consists in placing the body in various positions that would lead to the discipline of the bodily system. *Asana* helps the body to get rid of restlessness that blocks deep reflection and concentration. Posture brings the entire physical organism of the aspirant under the control of his will. It is different from other types of physical exercises, as its aim is to make the body best fitted for the highest type of experience.

The *hathayog in* keeps his body free from all impurities, his nervous system in tact and gains control over the different muscles of the body by the practice of the *asanas*. Speaking of posture Shankara says that the real posture is one in which the meditation on *Brahman* flows spontaneously and unceasingly.

He did not approve any posture that brings pains and thus distracts the mind from meditation on *Brahman*. Thus, for Shankara, the posture that a true aspirant of the spiritual path must learn and practice is that which genuinely helps meditation in the highest level, by bringing a serene, calm and relaxed state of the body. Shankara mentions two of such *yogic* postures that are truly conducive to the highest level of meditation, *viz.*, *Siddhasana* and *Mulabandha*. The former is a posture, which can help the aspirant to have a depth level concentration and meditation that can bring about complete merger with *Brahman*. The latter is a posture that can assist the aspirant to open his mind so totally to *Brahman*, that it can take hold of the mind of the aspirant, in the process restraining the mind from all other pushes and pulls.

The practice of the above mentioned *asanas* bring about an equipoise of all the limps of the body (*dehasamya*), which is conducive to the experience of total absorption in the *Brahman*.

Thus, for Shankara, while seated in meditation one should definitely use postures, but one's main focus should not be centred on posture. A mere turning of attention on posture alone by straightening the body is compared to a dried up tree, which does not bring about equipoise of limps. Therefore, a person should use postures that would open the person's whole attention on meditation on *Brahman*. In other words, practice of posture as an end is of no avail; but it must be practiced only in view of genuine help for meditation.

Breath-control: Genuine practice of the posture helps the aspirant to the practice of the second limp, breath-control (*pranayama*). It is a method with the help of which, the aspirant controls vital power of breathing, which is the basis of organic life. It keeps under check one's inhalation and exhalation. For Patanjali, the founder of the Yoga System, *pranayama* involves controlling the motion of inhalation and exhalation. There are three steps in breath-control.

The first one is *puraka*, which consists in taking in breath; the second is *kumbahaka*, which means to hold the breath taken in for some time in the lungs; and the last step is *rechaka*, which is to throw out the breath retained in the lungs during the second stage.

It was Patanjali's opinion that the mind would be naturally controlled if one practices the restraining of breath and thereby prevents mind's communication with the external world. But, Shankara does not subscribe to Patanjali's view. He holds that the breath is entirely dependent on the mind and not the mind on breathing. Therefore, by restricting the breathing one cannot restrain the mind. On the contrary, the control of the mind would effectively bring about restraining of the breath. So the aspirant, instead of wasting his energy in his attempt to restrain the breath, must always try to control the mind.

When the mind is controlled, it would automatically lead to the restraining of the breath. Thus, Shankara holds for the superiority of the mind over breathing and not vice versa. This makes him give the following definition of *pranayama*. "The restraint of all modifications of the mind by regarding all mental states like *chitta* as *Brahman* alone, is called *pranayama*."

Thus, genuine practice of breath-control is not a mere breathing in of air, retaining of air in the lungs and breathing out of the retained air. If this were so, *pranayama* would be a function done in relation to the physical sense organ of breathing, *viz.*, the nose.

But, for Shankara, breath-control has to do with the control of the mind in every aspect and meditation on *Brahman*. That is why he says the following regarding *pranayama*: "The negation of the phenomenal world is known as *rechaka* (breathing out), the thought `I am verily *Brahman*' is called *puraka* (breathing in) and the steadiness of that thought thereafter is called *kumbhaka* (restraining the breath). This is the real course of *pranayama* for the enlightened, whereas the ignorant only torture the nose."

Thus, for an Advaitin, *pranayama* if practiced as advocated by Shankara, can lead to complete control over the modifications of the mind (*chitta-vritti*). It helps him to achieve complete will power over one's life force. If practiced consistently, one can learn to restrain the vital process completely. According to Advaitin, *pranayama* has two purposes. Firstly, it brings about perfection in the body. Secondly, it helps to awaken the *pranic* dynamism, *i.e.*, vast stored up energy, which opens the aspirant for extraordinary consciousness.

As a result, the moral and spiritual possibilities are awakened due to the practice of *pranayama*. Breath-control, thus, helps the aspirant to control instincts, passions and impulses that disturb the peace of mind. It can be practiced in two ways. Firstly, as a purely physical practice, to keep the heart beat under control, without any plan of the control of the mind. Secondly, as a psycho physical practice to control the lower passions, to open in the seeker higher qualities and to help mental and spiritual discipline. The practice of *pranayama* in the latter way is of great help to the seeker of *Brahmanubava*.

When one has completed the practice of *Hathayoga*, he has made the first step into the process of *Brahmajijnasa*. The total control of body, by way of posture and breath-control prepares the seeker of *Brahmanubhava* to move towards the next stage of moral preparation.

Moral Preparation

In this section we would begin by pointing out the need, on the part of the aspirant, for moral transformation, in order that he can move on to the next step of *Brahmajijnasa*, *viz.*, the study of scriptures. Then we proceed to the content of the moral preparation, *i.e.*, the practice of four disciplines known as the instruments of spiritual path and the four qualities that the seeker of *Brahman* should acquire in order to be worthy of self-realization. Besides, we would also look into a few other requirements that the aspirant must practice on his way to *Brahmanubhava*.

Need for Moral Preparation: If the intellect is able to understand the import of the scriptural statements, it must be pure. The knowledge of *Brahman* revealed in the scripture, though expressed in terms of duality, still is the highest knowledge that can be known by the human intellect in the empirical realm. This knowledge cannot be grasped if the intellect is not open to understand eternal truth. Just as a stained mirror does not reflect things clearly, so is an impure mind cannot grasp *Brahman* intellectually from the study of the scriptures. Shankara affirms this point in his commentary on Mundaka Upanishad Bhashya as follows:

> *Though the intellect in all beings, is intrinsically able to make the self known [from the study of the scripture], still being polluted by such blemishes as attachment to external objects, etc., it becomes agitated and impure, and does not, like stained mirror or ruffled water make the reality of the self known, though it is ever at hand.Shankara uses another analogy to illustrate the same fact. Fire, by its nature, is able to burn the wood. But if the wood is wet, fire is not able to burn it. In the same way intellect, though it is able to know and understand the import of the scriptural statements, because it is clouded by passions and attachments to things, does not grasp* Brahman *from the study of the* mahavakyas. *In other words, scriptural knowledge fails to accomplish its end, if man is not perfectly pure of heart. The capacity of the mind to discriminate between*

truth and untruth is weakened, when passions and sensual pleasures sway the mind. Thus, it is important to prepare oneself spiritually and to free the intellect from all passions, attachments and prejudices, in order to know the true nature of Brahman *from the study of the scriptures and from the instructions of the* Guru. *So it is important that an aspirant goes through a course of moral preparation before he makes a serious attempt to study the scripture. Therefore, Shankara proposes four disciplines called the instruments of spiritual knowledge (*sadhanachatushtaya*), which the aspirant must practice before he ever begins the next state of the process of* Brahmajijnasa, *by the study of the scripture, under the guidance of the* Guru. *Now, we could turn out attention to the study of these four disciplines.*

Four Instruments of Spiritual Path: The four disciplines are aimed at preparing the inner personality of the aspirant, that he would be able to grasp and take in the truth found in the scriptural texts. It the aspirant works on these disciplines, he would be able to get the best out of the study of the scripture. The greater an aspirant is rooted in these practices the greater his understanding of the *mahavakyas*. Thus, the proposed disciplines are not aimed at frightening the beginner in the spiritual path, but to help him in pursuit of spiritual knowledge.We could now briefly speak of each of these disciplines of spiritual path in the following section.

Discipline in the Physical Level: The first discipline is the destruction of sins through the practice of austerities (*tapobhih ksinapapanam*), which is a discipline at the physical level. The practice of austerities (*tapas*) does not consist in giving oneself to self-torturing, which would frustrate the true spirit of the aspirant. But, rather, it envisages two types of disciplines at the physical level, *viz.*, a negative discipline which aims at preventing sensual dissipation by exercising self-control and a positive one that utilizes the conserved energy to move towards the highest ideal of self-realization.

These two disciplines correspond to the first two limps of *rajayoga* of Patanjali, the founder of the Yoga System. The first limp of *rajayoga* is *yama*, which consists of five general disciplines of self-purification. They are the following. Non-injury, which involves not harming others by thoughts, words and deeds. Non-lying, which consists in maintaining identity between thoughts, words and deeds. Non-stealing, *i.e.*, giving up the desire for other's things. Continence, *i.e.*, abstaining from sexual intercourse and lustful thoughts. Non-acceptance of gifts which would likely to stand in the way of meditation. The second limp of *rajayoga*, *niyama* consists in the practice of five virtues. They are the following. Cleanliness consists in the purity of body and mind. Contentment is the acceptance of what comes in life joyfully. Austerity is the practice of mortification. Study of the scriptures implies the repetition of the *mahavakyas* and the eternal syllable '*Aum*'. Worship of God involves knowledge of and loving devotion to God.

Shankara says that the negative discipline of *yama* and the positive discipline of *niyama* should not be practiced mechanically, but must be done by means of the knowledge that everything is *Brahman*. If practiced in this manner most of these negative restraints and positive practices will truly help the aspirant. When these two disciplines are applied not mechanically, but intelligently with the thought of *Brahman*, though their practice may involve pain, they would bring satisfaction and contentment. Besides, they free the aspirant from sin (*papa*) and the activities, which produce agitation of the mind and destroy mental peace. Thus, the first spiritual discipline, the discipline in the physical level, with its positive and negative practices, destroys sins and inclination to sin in the aspirant and generates in him purity of body and mind.

Discipline in the Mental Level: The second spiritual discipline aims at restraining the aspirant in the level of the mind. It is aimed at bringing in the seeker a calmness of mind (*santanam*). The mind, by its very nature, has the tendency to seek objects of the world through the help of the senses. The mind in seeking and knowing the objects goes out through the senses and modifies itself in the form of the object. It is in relation to these modifications of the mind (*chitta-vrittis*) it knows the objects. In doing so the mind remembers past enjoyments and anticipates in thought future possibilities. These, in turn, cause agitation in the mind, which prevents it from focusing itself on higher knowledge.

The discipline at the level of the mind is aimed at disciplining the mind from drifting into the enjoyments of past objects and ruminating about future joys. It fundamentally consists in not allowing the mind to get lost in objects, whether past, present or future by removing its attention from the senses which are mind's gate ways to the world of objects, thereby attaining mental peace and tranquillity. This mental discipline is comparable to the fifth limp of *rajayoga*, *viz.*, *pratyahara* (withdrawal of senses), which consists in removing the focus of the mind's attention from the senses. In other words, in *pratyahara*, the aspirant is helped to withdraw the sense organs from the respective objects, which happens only when the mind does not allow it to be modified into objects by shutting down the senses.

Shankara says that such a withdrawal would be complete only when the aspirant realizes that *Atman* is in all objects, allows this thought to envelop the mind and let the mind be absorbed in the supreme consciousness. In this manner, the turning of the mind from the senses and sense objects, and turning it towards higher levels of being makes the mind to attain the calmness of mind (*santanam*). That which helps the attainment of mental equipoise and calmness is the practice of silence (*mauna*). To quote Shankara: "The wise should always be one with the silence where from words together with the mind turn back without reaching it." What Shankara means by this statement is that silence is where *Atman* dwells, *i.e.*, in the depths of one's inner silence dwells the true self and a mind crowded with words will never be able to reach it.

The more the mind withdraws itself from words that are centred on the senses and experiences silence (*maunam*) it would arrive at the experience of *Atman*. Shankara says further: "Who can describe That (*Brahman*) whence words turn away? (So silence is inevitable while describing *Brahman*). [Therefore] silence is known among sages as congenital [*i.e.*, inseparable from *Brahman*]. [This is why] the observance of silence by restraining speech is ordained by the teachers of *Brahman* for the ignorant." Thus, for Shankara, the second spiritual discipline, aimed at attaining calmness of mind, involves both the silencing of the senses, by silencing the modifications of the mind in all its aspects. Therefore, the more an aspirant allows silence, which, according to Shankara, is congenital with *Brahman*, in his life, to that extent he would achieve mental poise and peacefulness.

Discipline in the Intellectual Level: The third spiritual discipline is one that focuses on the intellectual level, aims at freeing the aspirant from all desires (*viitaraginam*). A desire, is nothing else, but a plan entertained by the intellect to set right a lack felt within a person. When a person experiences an imperfection within himself, he sets in motion a scheme of action, prompted by the desire to experience a sense of well-being in the area of his lack. As long as a person is intellectually involved in accomplishing the plan that stems from desire, his mind remains agitated and disturbed. The imperfection experienced by the person often relates to the realm of his body, mind and intellect.

The person experiences this lack, and the desire to rectify this lack by a plan of action remains only as long as he identifies himself with these mental faculties, which he believes are the source of the imperfection. Thus, a person's identification of himself with the limited faculties of the body, mind and intellect, is the basis of all desires. On the other hand, if a person removes his focus of attention from these faculties and sees his life in relation to the supreme *Brahman*, then he would not perceive himself as identical with the limited mind-body organism. In doing so a person removes himself from the source of all imperfections experienced by him.

As a result, he does not any more experience imperfections, for he knows that his real being is all-perfect and eternal *Atman*. When this thought takes hold of a person, not only does he experience imperfections, but also sheds all desires, which emerge from the realization of imperfection within himself. Thus, withdrawing one's attention from the material layers of one's personality and concentrating on *Brahman* would effect the cessation of all imperfections in a person, and so bring about the end of all desires that disturb and agitate the mind.

This discipline at the intellectual levels is similar to the sixth limp of *rajayoga*, *viz.*, *dharana* (concentration or one pointed attention). It consists in holding the mind on to some particular object. In other words, *dharana* is the total concentration, in which the mind's focus is fixed fully on to an object. Shankara commenting on this point says that when the mind is fully concentrated on an object and realizes it as *Brahman* discards all names and forms superimposed on the object by ignorance, then alone a person is said to have reached the culmination of *dharana*. To quote Shankara: "The steadiness of the mind through realization of *Brahman* wherever the mind goes, is known as supreme *dharana*."

According to Shankara, one can attain this state of concentration by genuine practice of solitude and focusing on *Brahman*, who is beyond temporal distinction. "Solitude is known as space wherein the universe does not exist in the beginning, end or middle, but whereby it is pervaded at all times." When an aspirant enters the state of solitude, he becomes *Brahman*, who alone is solitary, since it admits no second at any time. Thus, by practicing solitude, the aspirant is able to withdraw his attention from everything, including the distinction of time, *viz.*, beginning, middle and end, and concentrate on non-dual *Brahman*, in whom resides the totality of time (*kala*). When an aspirant disciplines his intellect so as to arrive at such a state of deep concentration, wherein he is one with the solitary *Brahman*, all the desires that emerge from the experience of imperfection fall apart and the aspirant possesses compete well being and peace of mind.

Discipline in the Spiritual Level

The fourth spiritual discipline is a discipline at the spiritual level. It is the consequence of the first three disciplines, *viz.*, disciplines of the body, the mind and the intellect. It aims at creating in the aspirant a yearning for liberation (*mumukshuunam*). In this discipline the mental energy generated by the disciplines of the body, the mind and the intellect is conserved and directed for gaining liberation (*moksha*). It involves the rejection of all the worldly desires and substituting this rejection with the desire for gaining spiritual liberation. This discipline is facilitated by what Shankara calls the *drik-sthiti* (fixing the vision).

It consists in directing the aspirant's vision on *Brahman* alone, which is pure consciousness, and wherein ceases all distinctions of the seer, the sight and the seen. In other words, one must convert one's everyday vision of the world into one of knowledge, in which one views as *Brahman*. Shankara calls this the noblest vision, because in it there is no distinction of high or low, great or small, since everything is merged in *Brahman*. The fixing of vision on *Brahman* in the fourth discipline is possible because by way of the first three disciplines, the aspirant is purified in every aspect of his personality. Therefore, he can now direct all his energies to fix his vision on *Brahman* and desire his oneness with *Brahman*.

Thus, the fixing of the vision on *Brahman* makes the aspirant to let go all worldly desires and desire nothing else but liberation. In this manner a total orientation towards *Brahman* is brought about. These four spiritual disciplines—the disciplines at the physical, mental, intellectual and spiritual levels—help the aspirant move from the world to *Brahman* and to substitute one's worldly desires with desire for liberation. In going through these four disciplines, genuine moral preparation in the aspirant happens. As the result of these practices the aspirant becomes a qualified person (*adhikari*) to begin the study of the scriptural texts. In the process the moral preparation of the *adhikari* acquires four qualities, to which we turn our attention in the next section.

Four Qualities of the Adhikari

The qualified person acquires, as the result of the four disciplines called the instruments of spiritual knowledge (*sadhanachatushtaya*), four qualities. They are the following. The discrimination between the real and the unreal (*viveeka*); detachment from the unreal, *i.e.*, renunciation (*vairagya*); the practice of six fold virtues (*shadsampatti*)—calmness (*shama*), self-control (*dama*), self-settledness (*uparati*), forbearance (*titiksha*), faith (*shraddha*) and complete concentration (*samadhana*); and hunger for self-realization (*mumukshvata*). Each of these qualities morally prepares the student for the study of the scriptures and, therefore, they can be considered, as moral conditions required of the student, before ever he can undertake the deliberate and serious study of the scriptures. We could briefly describe each of them in the following sections.

Discrimination: The first moral condition required of the *adhikari* that he can begin his move towards the study of the scriptures is discrimination between the eternal and the non-eternal (*nitya anitya vastu viveeka*). It consists in an intuitive and firm conviction that *Brahman* alone is the absolute ground of all things, that which is really real and that all other things are unreal and phenomenal. According to *Vedantasara* it "consists of the discrimination that `*Brahman* alone is permanent substance' [unlimited by space and time] and that all things other than It, are transient." Discrimination is the most fundamental quality that is necessary in an individual, because without it the other moral conditions are not possible. In other words, if one has not discriminated between the real and the unreal and has not recognized the absoluteness of *Brahman* there is no motivation to strive for its attainment.

Renunciation: The second moral quality needed for the removal of ignorance is renunciation (*vairagya*). It consists in not seeking the enjoyments of the fruits here and hereafter. Shankara defines renunciation as follows: "The indifference with which one treats the excreta of a crow—such an indifference all objects of enjoyment from the realm of *Brahma* the realm of gods] to this world (in view of their perishable nature) is verily called pure *vairagya*." If a person renounces the enjoyment of this world, in expectation of better enjoyment in the next life, it would be a renunciation tainted with desires.

Such a renunciation is not a true renunciation and it would not open the door for true knowledge. But a renunciation that comes out of the deliberation on the passing nature of this world and undertaken in recognition of the ultimate truth of *Brahman* is pure renunciation that would lead to higher levels of knowledge. Thus, genuine practice of renunciation implies, firstly the negation of one's little self (*jiva*) and secondly the assertion of the reality of *Atman*. When a person gives up the ego's claims of `I', `me', `my' and `mine', thereby rejects egoistic feelings that engender one's false personality and asserts that his ultimate self is the universal spirit, a person truly practices renunciation. Such a renunciation truly leads a person to knowledge. The practice of such renunciation implies rigorous self-discipline to control the inner tendency of the aspirant to wander amidst sensuous experience and enjoy it. Renunciation is aimed at purification of emotions, perfection of the mental being and mastery over thought-consciousness. It would enable the aspirant to eliminate lawless tendencies, impulses and tendencies of animal nature, and build up good habits of mind. Renunciation is, thus, the giving up of all the pleasures of the eyes, the ears, and the other senses. By renunciation one also gives up objects of transitory enjoyment, the desire for the body, as well as for the highest kind of spirit-body of a god.

It is an attitude indifference to all the seen and unseen results attainable by various means, whereby one becomes devoted to the scriptural teaching, and not carried away by one's own natural desires. But renunciation is not the state of absolute desirelessness or *Brahmanubhava*. Rather it is a state where one desires nothing but self-realization. In renunciation one forgoes the pleasures the world can offer in order to concentrate on the study of the scriptures. Therefore, a life of renunciation gives the aspirant the opportunity to go for uninterrupted search after the goal of human existence, by a deeper understanding of the *mahavakyas*. Any person who is interested in absolute freedom must take every step to transcend earthly ties and involvement, by way of renunciation. Vedantins speak of two types of renunciation, *viz.*, the *vibidisha sannyasa* and the *vidwat sannyasa*.

The *Vibidisha sannyasa* is the renunciation for the acquisition of knowledge. It is the renunciation practiced by the seeker. It does not merely consist in giving up worldly pleasures, but rather involves systematic seeking of the life of wisdom. The aspirant while practicing this type of renunciation, besides denouncing attachment to the phenomenal enjoyments seeks after hearing and reflecting about *Atman* and concentrating upon it.

The *Vidwat sannyasa* is a renunciation one practices after the attainment of true knowledge. Complete freedom from all desires and activities is characteristic of this type of renunciation. Thus, it is an existence without desire, in which renunciation coexists with true knowledge.

Practice of Six Treasures

Thirdly the aspirant should try to live a virtuous life by the practice of six treasures (*shadsampatti*). When practiced, these virtues will help the aspirant's inner faculties and make the cultivation of higher knowledge possible. The six virtues are the following. Firstly, calmness (*shama*), which consists in developing the quality of inner serenity, that one can dwell on *Brahman* after abandoning all the desires by renunciation. Secondly, self-control (*dama*), which helps the aspirant to restrain his senses of sight, hearing, touch, smell and taste, and thus open him to hear and listen to the instructions of the teacher. Thirdly, self-settledness (*uparati*) is the withdrawal of all forms of enjoyments. It helps the aspirant to relinquish all the joys of worldly life and embrace the life of a monk (*sannyasin*).

Apparently *uparati* differs very little from *shama* and *dama*. But there is a real difference between the former and the latter two. In practicing *shama* and *dama* there is an effort to restrain the mind's outgoing propensities, such as, curbing the mind's attention from all objects except hearing the scriptures, thinking of their meaning and meditating on them, and restraining external sense organs respectively.

But in *uparati* the equipoise of the mind becomes spontaneous and no effort is made to gain it. Fourthly, forbearance (*titiksha*) is the endurance of all sufferings in life. It helps the aspirant not to be agitated by love and hate, pleasure and pain and all such pairs of opposites. Fifthly, faith (*shraddha*) the firm conviction and intuitive belief in the existence of the ultimate reality. At the same it is not a mechanical or unquestioning belief, but a conviction based on intellectual understanding of what is taught in the scriptures about the ultimate reality. Sixthly, complete concentration (*samadhana*) is the fixing of the mind on *Brahman* as taught by the scripture and competent teacher. These six virtues direct one's attention to the attainment of *Brhamanubhava*.

Hunger for Liberation

The fourth moral quality of the *adhikari* is his hunger for self-realization (*mumukshvata*). It consists in possessing an intense desire to get rid of ignorance and to attain *Brahmanubhava*. "It is an intense longing of the student to free himself from all bondage pertaining to the body, the mind and the ego." It is not a restless desire, but a result of the ethical practices mentioned above. Without this longing for liberation the ethical practices are really meaningless because a man may become morally perfect and may achieve many supernatural powers, but if the desire for self-realization is absent in him, all his virtues will be of no significance. It is the longing and desire for true knowledge or self-realization that gives motivation and meaning to the whole process of *Brahmajijnasa*.

Other Requirements of an Adhikari: We have looked into the four disciplines of the spiritual path and the four qualities required of a qualified student to undertake the study of the scriptures, in the last two sections. In the present section, we could move on to consider some other general requirements demanded of an *adhikari* that would also facilitate the scriptural study.

The aspirant can begin his study only if he practiced a life of chastity (*brahmacharya*) and other austerities of student life prescribed by the Vedas. Besides, he also must acquire knowledge of the contents of the Vedangas, books auxiliary to the Vedas. The Vedangas include the science of proper articulation and pronunciation (*shiksha*), etymological explanation of difficult Vedic words (*nirukta*), the science of prosody (*chhandas*) astronomy (*jyotisha*) and ritual and ceremonies (*kalpa*). Regarding the last of the Vedangas, *viz.*, ceremonies, there are a number of them which the *adhikari* is expected to practice or avoid practicing. The aspirant (*pramata*) is expected to avoid the practice of *kamya karma*, which are ceremonies performed with a definite motive or desire, such as, *jyotishtoma*, which enables the performer to get the desired fruit like living in heaven.

He is also asked to refrain from performing *nishidda karma* (the forbidden acts), for they bring undesired results, such as, going to hell. These forbidden acts include actions like slaying a *Brahmin*, drinking and other similar vices. The qualified student must practice *nitya karma*, which include *sandhya vandana*, such as, morning, noon and evening prayers, which are obligatory for the three higher castes. The non-performance of *nitya karma* though may not constitute a new sin, but will weaken a person's will, thereby prevent his propensity to check the inclinations to commit new sins. Besides, the seeker of the scriptural study must practice *naimittika karma*, which are rites performed on special occasions.

The aspirant also must practice rites of penance (*prayaschittas*), such as *chandrayana*. He is also expected to practice devotions (*upasanas*). The practice of *nitya karma* and similar other works are aimed at purifying the mind, by way of destruction of sin and acquisition of virtue. The purification of the mind effects in the aspirant an awareness of the nature of the phenomenal existence (*samsara*). From the recognition of the peripheral nature of *samsara* results renunciation (*vairagya*), which gives rise to desire for liberation. The desire for liberation gives the incentive to search for the means of attaining it. From this search for the means arises the practice of *yoga*, which leads to the habitual tendency of the mind to concentrate in the knowledge of the meaning of the *mahavakyas*, such as 'That art Thou', which destroys ignorance and identity of the self with *Brahman* is experienced.

Thus, the regular practice of *nitya karma* other similar practices lead the seeker to the highest knowledge. The *upasanas* aim at helping the aspirant in concentration of the mind on the *Ishtadeevata* and deepen it by loving devotion. In this manner all such practices help the *adhikari* to build up genuine openness to the inner self and universal *Brahman*, thereby facilitates the study of the scripture. While the aspirant goes through the four disciplines, practices the four qualities and performs other practices, he must never forget the fact that self-realization cannot be attained by the direct and mechanical performance of these spiritual practices.

While not undermining the value and efficacy of these practices, he must remember that these disciplines are only preliminary preparations necessary for gaining ultimate experience of self-realization and the final removal of ignorance is effected only at the dawn of true knowledge. This can be explained with the help the illustration of the role of fire in cooking. Before fire is used to cook the food, the raw food is prepared by way of cleaning, peeling and cutting. These are necessary preparations in the cooking process. The real cooking is done when fire is applied on the food. It is fire that really and in actual fact cooks the food. In the same way, it is knowledge of the self that leads to final liberation, even though all other spiritual disciplines the *adhikari* has to go through are auxiliary preparations leading to the dawn of knowledge. It is very significant that the aspirant constantly remembers this truth, through out the process of *Brahmajijnasa*, the path to self-realization.

We have looked into the various aspects of the moral preparation the aspirant has to go through in order that he becomes fully qualified to move into the next stage of *Brahmajijnasa*, the intellectual preparation. Intellectual preparation consists in the study of the import of the *mahavakyas*, under the guidance of the *Guru*. The topic of the intellectual preparation occupies our attention, as we move into the next section.

Intellectual Preparation: The aspirant who is endowed with the above mentioned disciplines of spiritual knowledge is qualified to undertake the actual study of the scriptures which will, in turn, bring about the actual removal of ignorance. Shankara says: "This knowledge [that the self is *Brahman*] should be imparted only to him whose mind has been pacified, who has controlled his senses and is freed from all defects, who has practiced the duties enjoined by the scriptures and is possessed of good qualities, who is always obedient to the teacher and aspires only after liberation and nothing else." The moral disciplines purify the intellect of the aspirant, freeing him from all passions and attachment, so that he can give himself uninterruptedly to the study of the scriptures.

More than other two stages of *Brahmajijnasa*, in the stage of intellectual preparation the role of the *Guru* is very significant. Fully qualified student cannot venture into the vast ocean of scriptural learning, unless competent *Guru* leads him into it. Even a person, who is well versed in the scriptures, should not undertake the journey towards *Brahman* without the guidance of a *Guru*. Therefore, for Shankara, instructions of the Guru are absolutely necessary, for the knowledge of *Brahman*, whether the aspirant is a person with scriptural knowledge or a qualified beginner (*adhikari*). Shankara affirms this in the Mundaka Upanishad Bhashya as follows: "One though versed in scriptures, should not search independently after the knowledge of *Brahman*."

What Shankara implies by this statement is that every seeker of knowledge of *Brahman* must sit at the feet of the proper and competent *Guru*. Now we must clarify the nature of a proper and competent *Guru*. For Shankara, the *Guru* is a spiritual guide, who is learned in the Vedas, desireless and sinless. He says in his commentary on Brihdaranyaka Upanishad that the *Guru* is "one who is learned in the Vedas, without sin and not overcome by desire." The genuine *Guru* is one who is well versed in the study of the Vedas and who has knowledge of the true import of the meaning of the Vedantic aphorisms.

Thus, the *Guru* is a true teacher of knowledge of *Brahman* to any one who wishes to make the journey towards *Brahman*. Other than these qualities, there is one quality that is most significant for a *Guru* being genuine is that he lives entirely in *Brahman* (*Brahmanishtha*). Shankara explaining this word `*Brahamanishtha*' says that the *Guru* is one who, after renouncing all forms of *karma*, is centred on *Brahman*, devoid of all attributes and one without a second. In other words, the true teacher is one who has experienced the identity between his self and *Brahman*. Shankara says that the seeker of knowledge of *Brahman* should approach such a *Guru* in the spirit of humility and service, and with suitable gifts in his hands.

When a student approaches the *Guru* in this manner, the *Guru* through his infinite grace instructs the pupil by the method of *apavada* (de-superimposition) and removes all superimposition (*adhyaropa*) on *Brahman*, thus, prepares him for the knowledge of *Brahman*. Mundaka Upanishad states: "To that pupil who has approached him [*Guru*] with due courtesy, whose mind has become perfectly calm, and who has control over his senses, the wise teacher should truly impart the knowledge of *Brahman* through which he knows the Being, imperishable and real."

Thus, begins the intellectual preparation, the third stage of *Brahmajijnasa*, for the qualified student, at the feet of the *Guru*. The study of the scripture and the understanding its import and meaning takes place in three stages. The first stage is hearing (*sravana*), which, is followed by the state of reflection (*manana*) and the final state of *Brahmajijnasa* is meditation (*nidhdhyasana*). These three, namely, hearing, reflection and meditation, constitute the objective intellectual conditions for the removal of ignorance. Now we could proceed to an elaborate study of each of these stages of intellectual preparation.

Hearing

In this section we would briefly clarify the meaning of *sravana*. We would also make an attempt to give an illustration with the help of which its meaning would be made clear.

Meaning of Hearing: Hearing implies the idea of being taught. At the first stage of understanding the meaning of the Vedantic statements, the competent teacher introduces the aspirant to the teachings of Advaita. The need for a teacher at this stage is very much stressed in the scriptures. "A *Guru* is like a boat on the boundless ocean which has for its water the principal struggle due to the rotation of the cycle of birth, decay and death." Just as a boat is a place of safety for one who struggles in the ocean aimlessly, so the teacher, by his teaching directs one to union with *Brahman*, freeing him from transmigration. Again Katha Upanishad Bhashya states: "When propounded by a teacher who sees no variety and is one with (*ananya*) the propounded *Brahman*, doubts whether the self exist or not, — does not arise any longer for the self is such that it banishes doubts of all kinds." Thus, *sravana* is the initiation of the aspirant to the traditional Vedantic doctrine transmitted and passed on by the teachers.

Sravana at the same time is the mental activity, which helps the understanding of the upanishadic texts, leading to their only import, *i.e.*, *Brahman*. In other words, *sravana* is not a mere hearing the truth about *Brahman* from the teacher or from the scriptures in a blind manner. But it involves an ascertaining on the part of the aspirant on what is heard. Thus, in *sravana* the student ascertains and establishes the true import of the scriptures, *viz.*, '*Brahman* is one without a second'. Without a genuine ascertaining *sravana* would be fruitless. This ascertaining is achieved by an examination of the texts through six tests or characteristic signs, namely, commencement and ending, repetition, uniqueness, result, eulogy and reason. We could briefly clarify each of these texts.

The first is commencement and ending. Commencement involves the presentation of the subject matter to be taught to the student, in *sravana*, at the beginning of a section, that the aspirant clearly comes to know the topic of his study. In the ending the same truth is restated not as a hypothesis but as a verified fact that the student knows about it with certainty. The second test, the repetition consists in frequent presentation of the subject matter in different parts of the section. It is aimed at helping the student to become more and more aware of the import of the subject matter by ascertaining each time it is repeated. The third is uniqueness of the subject matter. It consists in that the subject matter of a section is not available through any other source of knowledge, but only has to be understood in relation to the study of the scriptures.

The fourth is the result. It is the utility of the subject matter of the section for the qualified student to move into the higher stages of *Brahmajijnasa*. In other words, the result consists in the utility of the subject matter, for its attainment. The fifth test, eulogy consists in praising the subject matter in different places of the section. It is aimed at instilling in the student a desire to hear attentively and inculcate in him its significance. The last test is reason, which consists in demonstrating the subject matter with rational arguments. It would help the qualified student to understand the thinking behind the subject matter and understand its import, which in the last analysis would lead to the right ascertaining of the meaning of the subject matter. By the use of these six tests the student makes a deep effort to understand the import of the subject matter that he heard from the teacher or read from the Vedas. We move on, in the next section, to give an illustration of *sravana*, which would further clarify its meaning.

An Illustration of Hearing: In the sixth chapter of Chandogya Upanishad we have a typical illustration of this first stage, namely hearing. Here the aspirant is Sivetaketu the grandson of Aruna. His father instructs him by using the six texts of *Sravana*. The subject matter of this instruction is '*Brahman*, the one without a second'. The analysis of the six tests used in hearing, in relation to the text in question, would give us an idea about the practice of *sravana*.

The first test is commencement and ending. They refer to the presentation of the subject matter of the section at the beginning and at the end of the section. The subject matter is introduced in the beginning of the section in the words, "one without a second" and again at the end in the words, "thus has all this world, that [*Brahman*] for its self." The second test is repetition, which consists in the frequent presentation of the subject matter again and again. For instance, the subject matter '*Brahman* is one without a second' is repeated nine times in this section, in different words, especially through the *mahavakya* 'That art Thou'.

The third is uniqueness, which means that the nature of the subject matter is such that it cannot be attained through any other means of knowledge than the study of the scriptures. In this section the subject matter '*Brahman* is one without a second' cannot be achieved in any other way except from the understanding of the meaning of the *mahavakyas*. This is indicated in the words: "Verily, those venerable men did not know this; for if they had known it, why would they have not told it to me? The implication of this statement is that `the knowledge of *Brahman*, without a second' can be achieved only from the scriptures.

The fourth test, the result, is the usefulness of the subject matter of the section. In this section the realization that *Brahman* is one without a second, brings about self-knowledge to the seeker. In other words, the utility of the knowledge of *Brahman* one without a second is its attainment. The fifth, eulogy, consists in praising the subject matter at different places of the section. It is found in this section in the words, "Have you ever asked for, that instruction by which one hears what has not been heard; one thinks what has not been known; one has spoken in praise of *Brahman* without a second". Finaliy reason, which consists in demonstrating the subject matter of the section. In this section we have the demonstration of `*Brahman* one without a second' in the words: "As by a lump of clay, all that is made of clay is known every modification being an effect of speech, a mode and the clay is the only reality." In other words, just as in knowing a lump of clay, we can know the nature of all things made of clay, so also in knowing that *Brahman* is one without a second, the nature of everything in the world is known, for *Brahman* is the ground of everything.

The examples of a nugget of gold and a pair of nail scissors are also used to demonstrate the truth of `*Brahman* is one without a second', in the same fashion as the lump of clay. In this manner three reasons are furnished to demonstrate the subject matter of the section, *viz.*, `*Brahman* is one without a second'. Thus, the aspirant hearing the explanation of the *mahavakya* from the teacher, become familiar with its import and ascertains its meaning.

Reflection

In the last section we clarified the meaning of *sravana* and how it is used to help the qualified student to ascertain the true import of the *mahavakyas*. In this section, we will elaborate the second stage of the intellectual preparation, *viz.*, reflection (*manana*). We do this by analyzing its meaning and method with the help of illustrations.

Meaning of Reflection: The Keena Upanishad Bhashya speaks about the second intellectual condition required for the removal of ignorance, *viz.*, *manana*, as follows:

> *After being addressed by the teacher, the disciple (shyshya) sat at the solitary place [vijana-desha] and attended to nothing else (ekanta), concentrated his thoughts (Samhita) and pondered over the meaning of agma [the traditional teaching pointed out by his Guru], arrived at a conclusion through reasoning, made it [the teacher's instructions] his own experience, went back to his teacher and explained: 'I think, I now know Brahman'.*

This passage clearly shows the nature and function of manana. It is a mental activity, which consists in the employment of favourable arguments for the removal of the apparent contradictions that might arise during the study of the scripture against other means of valid knowledge. The truth pointed out by the teacher is difficult to grasp and seems to contradict the ordinary perception and knowledge obtained from the pramanas other than the scripture.

Thus, it is very important that the aspirant strengthen his conviction at this stage, by looking for rational basis for the teaching received from the teacher in sravana. Prof. Ramamurthi clearly points out the role of manana as follows: "The purpose of it [manana] is to fortify one's conviction of the truth from the scripture and to rid oneself of all doubts.

Another important function of reflection is to make one comprehend the real meaning of the scriptural statements by consistently interpreting them so that the apparent inconsistencies are resolved." But, though at this stage of reflection, one looks for arguments and reasons to justify the knowledge of Brahman received in sravana, still the arguments and reasons sought to strengthen one's conviction must not be contradictory to the teaching of the Upanishads.

For, the scriptural authority is absolute, and reason is subservient to revelation, especially with regard to the knowledge of Brahman. *Therefore, for Shankara,* manana *is that continuous reflections performed with the aid of reasoning and that are subservient to the teaching of the Upanishads, upon the secondless reality of* Brahman, *that is known through* sravana. *Having clarified the meaning of* manana, *we could proceed to analyze the negative method used by the aspirant during the state of reflection, in order to de-superimpose phenomenal qualities from* Brahman.

The Method of Reflection: At the stage of *manana*, the aspirant makes use of the negative method of Advaita Vedanta, *viz.*, *apavada*, more than ever before. *Apavada* consists in the elimination of what some is not, in order that one may attain the truth about that particular thing. In other words, it is "the elimination of the falsely superimposed attributes (*vivartha*) in order to discover the true nature of a thing."

The rope appears as a snake in an illusion. By negating the illusory snake the true nature of the rope is perceived. By negating the attributes of the illusory water in the mirage one discovers the true nature of the desert. By negating the illusory silver that we obtain the true nature of shells scattered on the beach. This ability to distinguish between the real and the unreal, Shankara calls discrimination. It is this discriminating ability that helps us to understand the true nature of things. It is much the same with our knowledge of *Brahman*. By negating, through discrimination, the attributes of the non-self, one attains the true nature of *Atman* and by negating the world of names and forms one attains the knowledge of *Brahman*, the absolute reality.

Negation, then, consists in the refutation of the knowledge established by the method of superimposition (*adhyaropa*). In the scriptures we find many statements about the ultimate reality, expressed in terms of negation. "That which is invisible (*adneesya*), ungraspable (*agranya*), without family (*agootra*), without caste (*avarna*), without sight or hearing (*acaksisrotra*), without hand or foot (*apanipada*), immortal (*nitya*) imperishable (*avyaya*)" is *Brahman*. "The self is that which has been desired as neither, this nor that." "It (*Brahman*) is imperishable., undecaying., unattached., unsettled. It never feels pain, never suffers injury, it is transcendent." These passages from the scripture witness to the fact that these negated qualities like birth, death, hunger, thirst, pain and pleasure never did belong to the absolute *Brahman*, even though these were attributes given to *Brahman*, by one in the state of ignorance.

Neither do these superimposed qualities affect the ultimate reality in any way. Thus, the "*via negativa* of Advaita Vedanta safeguards the unqualified oneness of the state of being called *Brahman*." Brihadaranyaka Upanishads speaks of *Brahman* with words '*neti neti*' (not so, not so). In fact this statement does not make us perceive something direct as the statement 'this is a book' make us perceive a book. Therefore, the statement '*neti neti*' does not tell us anything positive about the nature of *Brahman*. But, in fact, it is the most proper way of describing *Brahman* since it is devoid of all qualities. The term '*neti*' consists of two words, namely, '*ne*' which means 'not' and '*iti*' which means 'so'. The word '*iti*' indicates the presence of something or a quality right here, which is negated by the word '*ne*'. Thus, '*iti*' used with '*ne*' points to something that is negated. The repetition of '*neti*' twice covers all possible predications that are to be eliminated.

The statement '*neti neti*' only denies the attributes superimposed on *Brahman*, but not the *Brahman*, for such denial of both *Brahman* and the qualities superimposed on *Brahman* would lead to pure void (*suunya*) and to Nihilism (*Suunyavada*). For Shankara says, "know that the *Sruti* 'not large' etc. is meant to negate the false superimposition (of largeness, smallness, etc., on the self) as it would be description of void if it were meant to negate those qualities from one other than the self." Just as the denial of the illusion of the snake leaves us with the reality of the rope, so too the denial of the qualities superimposed on *Brahman* reveals *Brahman* in its entirety. For according to Shankara '*neti neti*' denies not absolutely everything, but everything except *Brahman*.

At the same time the use of '*neti neti*' before a descriptive sentence does not necessarily mean that the particular descriptive sentence in question is false, but rather it only means that the sentence is not applicable or not appropriate in the case of *Brahman*. Consequently what this approach has done is to have generated a third kind of connotation which one can make of descriptions. No longer are descriptions either true or false; some of them may be a third kind of evaluation, namely, what had been termed 'inappropriate' or 'inapplicable'. Some times in scriptural passages we find twofold negation. In statements like *Brahman* is "not known and beyond unknown" and "neither gross nor subtle" what is denied is not only one attribute but also its opposite. Shankara comments on the statement, which describes *Brahman* as "neither *sat* (existent) nor *asat* (non-existent)" as follows. He says "since the Knowable (*Brahman* / *Atman*) is beyond the reach of the senses it cannot be an object of consciousness accompanied with the idea of either existence or of non-existence, and therefore, not said to be *sat* or *asat.*" *It* is clear from this comment of Shankara that in all such negations, the notion of *sat* or *asat* are understood form the *vyayahara*, and not *paramartha* point of view.

From the supreme point of view of absolute knowledge, the term '*sat*' applies only to *Brahman*. But it is quite natural that one understands, in the course of *Brahmajijnasa*, such terms as '*sat*' or '*asat*' in the ordinary meaning, they have for those who are still immersed in ignorance. In that case what the aspirant understands by `*sat*' primarily is the object of senses, which should be negated with respect to *Brahman*. Thus, the negative method completely does away with all false attribution of *Brahman*, and thereby paves the way for true knowledge.

Now that we have analyzed the meaning of negative method (*apavada*) used by the aspirant in the stage of *manana*, we could go on to give a few illustrations of this method and de-superimpose the superimposed phenomenal qualities on *Brahman* in the next section.

Illustrations of the Negative Method: Here, we attempt at two illustrations, in which the negative method is used to de-superimpose *Brahman*, the universal spirit and *Atman* the inner spirit in man. The twofold de-superimposition process clearly points to the identity between the universal spirit and the inner self. These illustrations also would give us an understanding of what an aspirant is expected to do in the state of *manana*, *viz.*, that he is called to form right convictions about the true nature of *Brahman* and *Atman* by way of de-superimposition.

We find the first illustration in the Brihadaranyaka Upanishad. It is a regress to *Brahman* by way of negation. Gargii, the daughter of Vacaknavii questions Yajnavalkya regarding the nature of *Brahman*. She begins from her experience of the phenomenal world, and moves towards *Brahman*, and at each stage Yajnavalkya uses negative method to counter her question. Her question proceeds as follows. Since the world is woven on water, on what the water woven? Yajnavalkya replies that it is on the wind that the water is woven. She asks further: 'On what the wind is woven?' to which the reply given is that the wind is woven by the atmosphere worlds. To her question 'on what the atmosphere worlds woven?' Yajnavalkya replies that they are woven by the worlds of the Gandharvas. To her plea 'on what the worlds of the Gandharvas woven?' he answers that they are woven by the worlds of the sun. 'On what the worlds of the sun woven? She asks and the reply is that the worlds of the sun are woven by the worlds of the moon. She questions him further saying 'on what the worlds of the moon woven?' and he replies that they are woven by the worlds of the stars.

To her inquiry 'on what the worlds of the stars woven?' Yajnavalkya replies that they are woven by the worlds of gods. She questions him again 'on what the worlds of gods woven?' and he satisfies her curiosity by saying that the worlds of the gods are woven in the worlds of Indra. To her question 'on what the worlds of Indra woven?' he says that they are woven on the worlds of Prajapathi. Again she asks 'on what the worlds of Prajapathi woven?' and he replies that the worlds of

Prajapathi are woven in the worlds of *Brahman*. Gargii persists in her questioning and asks 'on what the worlds of *Brahman* woven? Yajnavalkya replies, "Gargii, do not question too much, lest your head fall off. In truth you are questioning too much about a divinity about which further questions cannot be asked. Gargii do not over-question." Thus, Yajnavalkya using the method of negation moves towards *Brahman* eliminating all adjuncts of limitation (*upadhi*) starting from the lowest.

The second illustration of negative method is given in Sadananda Yogindra's book *Vedantasara*. Here the author attempts to de-superimpose the superimposed inner self (*Atman*), by bringing together different views held regarding the nature of the self. Each succeeding view is more refined than the preceding one as the former attempts to identify the self with higher faculties, while the latter identifies the self with lower faculties. In setting forth each of these views the author presents the threefold proofs, *viz.*, the scriptural evidence, inference and direct perception, given by the respective schools to justify its view of the self. At the end the author refutes all these fallacious views and sets forth the true nature of *Atman* as taught by Vedanta.

The first view represents the view of the most deluded common man, who identifies his self with an extraneous object like his son. The three reasons given in justification for this false identification are the following. The scriptural statement 'Verily the self is born as the son", the inferential fact that one loves one's son as one's own self and the experiential fact that one feels oneself prosperous or ruined according to one's son fares well or ill in life. This view of the self cannot be justified because one's self can never be his son, even though the son is very dear to oneself.

Four different schools of Indian Materialism (Charvakas) propound the next four views. The first school holds that the self is the physical body. They quote the *Sruti* statement "Man is constituted of the essence of food" as the scriptural evidence. From the fact that a person saves himself from a burning house, even leaving his son behind, they point out that a person's self is his body, which he wishes to save. The perceptual experience of oneself as `I am thin', `I am short', etc. also helps one to identify oneself with one's body. The second school of Materialism refutes the contention of the former school that body is the self by arguing that body become inert and inactive when sense organs cease to function.

Therefore, for them the sense organs is the self. The quote the scriptural passage `The sense organs went to their father prajapathi and said.", to justify their point of view. They infer that senses are the self from the fact that the movement of the body ceases to exist when sense organs cease to work. One's perceptual experience that `I am blind of one eye' and `I am deaf' clearly points to the fact that one's self is the sense organs.The third school argues against the previous school by saying that the activities of the senses are dependent on the vital force (*prana*). Therefore, for them vital force is the self. They give the scriptural passage "Different from and more internal than this (the physical body) is the self, which consists in vital force", to substantiate their view. From the fact that the cessation of the vital force leads to the cessation of activity of the sense organs, they infer that vital force is the self. The perceptual evidence that `I am hungry' and `I am thirsty' which are associated with vital force, also indicate that the activity of the body and the senses depend on the vital force, and so it is identical with the self.

The proponents of the fourth school of Materialism rebut the arguments of the third school on the ground that the activity of the vital force is controlled by the activity of the mind (*manas*). Thus for them mind is identical with the self. The *Sruti* passage they use to support their claim is that which says: "Different from and more internal than this (which consists of the vital force) is the self which consists of the mind." From the fact that vital force ceases to exist when the mind stops functioning they conclude that the mind is the self. Besides, the perceptual experience of `I am considering the pros and corns', also indicate that the mind, which has the capacity for weighing over the subject, is identical with the self.

The Buddhist Idealists (Yogacharas) contradict the view of the last school of Materialism. Since they consider the mind as a mere instrument of pleasure and pain, and that weighs the pros and corns, it needs an agent to act. This agent that controls the mind and acts on its behalf is the intellect (*buddhi*). Thus, for the Buddhist Idealists, the intellect, the agent of activity is the self. The scriptural passage they use to support their view is the following: "Different from and more internal than this is the self, which consists of consciousness." From the fact that mind as an instrument of judging becomes powerless at the absence of the agent intellect, they infer that *buddhi* is the self.

Besides, the perceptual knowledge expressed in statements, such as `I am the agent' and `I am the enjoyer', also indicate that the intellect is the self. The exponents of Miimamsa schools of Prabhakara and Kumarila Bhatta held the next two views regarding the nature of the self. The Prabhakaras refute the view of the Yogacharas saying that the intellect cannot be the self, as in the state of deep sleep the intellect and all other faculties merge in ignorance. Thus, for the Prabhakaras the ignorance, that characterizes the state of deep sleep, is the self. They substantiate this view with the help of the scriptural passage, "Different from and more internal than this [intellect] is the self which consists in bliss." From the fact that in deep sleep the body, the vital forces, the mind and the intellect merges in ignorance, they infer that ignorance is the self.

The perceptual experience of one as `I am ignorant' and `I am devoid of knowledge' clearly points to the fact ignorance is the self. Thus for Prabhakaras, as there is no consciousness in the state of deep sleep, it follows that the state of ignorance is the self. The Bhattas deny the view of the Prabhakaras, *viz.*, the ignorance is the self, on the ground that the in the state of deep sleep a person contains elements of both consciousness and unconsciousness.

Therefore, they claim that consciousness associated with ignorance is the self. Thus, for them, self is associated with knowledge and ignorance, consciousness and unconsciousness. They quote the scriptural statement "During dreamless sleep the *Atman* is undifferentiated consciousness", to substantiate their view. They also infer this truth from the fact that both consciousness and unconsciousness are present in the state of dreamless sleep. They explain this point as follows. Unless a man retains consciousness in the state of deep sleep, he cannot say on waking that he had slept well, because in deep sleep sense organs do not function and this knowledge that he had slept well cannot come from sense organs.

Therefore the self must possess consciousness as an inherent quality. At the same time there is also unconsciousness in the state of deep sleep, as one who goes through the experience of deep sleep fully unaware of what had happened during the period of sleep. Thus the Bhattas infer that the self is consciousness associated with ignorance. The perceptual experiences of `I had a sound sleep' and `I did not know anything then, *i.e.*, at the time of sleep' also indicate that consciousness associated with ignorance is the self. The Madhyamika school of Buddhism (Suunyavadins), interpreting the sayings of Buddha literally maintain that void is the self. They refute the Bhatta view, by saying that the self is neither consciousness, nor unconsciousness, but non-existence. They state the scriptural statement "In the beginning there was non-existence", in favour of their philosophical point. From the fact that there is an absence of everything in the state of deep sleep, they infer that there is nothing called self and that the reality is void.

They prove this point also from the perceptual experience of a man who has just awakened from deep sleep, which makes him say that `during the dreamless sleep I was non-existent'. All these eight different views of the self, present the self as one's son, one's body, one's vital force, one's mind, one's intellect, ignorance, consciousness associated with ignorance and non-existence. Each of these successive views negates the former by identifying the self with something higher than the previous knowledge of the self. So far eight fold negations have been made. The author of the book of *Vedantasara* states that all these negations have moved in the right direction, but they have not taken us to the true nature of the self. Having said this he makes an attempt to show the true nature of the self, by rebutting the arguments of these various schools. To quote him:

Since all these fallacious citations of scriptural passages, arguments and personal experiences, made by the different classes of people enumerated above beginning with the extremely deluded [common man's view of identifying his son with the self], in support of their respective views about the self, the subsequent view contradicts the previous one, it becomes quite clear that all these items from the son to void are not the self.

Moreover none of the items from son to void is the self, because all those fallacious citations of scriptural passages, arguments and personal experiences in support of them are all nullified for the following reasons: first because they contradict strong scriptural passages which describe the self as not gross, without eyes, without vital force, without mind, not an agent, but consciousness, pure intelligence and existence; secondly because they are material and are illumined by pure consciousness and as such are unreal, like a pot, etc.; and lastly because of the strong intuition of the man of realization that he is Brahman.

Therefore, the innermost consciousness which by nature eternal, pure, intelligent, free and real, and which is the illuminer of those unreal entities [such as son, body, vital force, mind, intellect, ignorance, consciousness associated with ignorance and void] is the self. This is the experience of the Vedantist.

This passage quoted from *Vedantasara*, clearly states the true nature of *Atman*, by way of the method of negation, in the process giving us a clear illustration of negation as a method to attain the knowledge of the absolute reality.

Thus, in the state of *manana* the aspirant de-superimposes, by way of negation, the phenomenal reality from *Brahman* and *Atman*, and understands the full import of the Vedantic statements like 'That art Thou', in their indirect and implicit meaning. *Manana*, therefore, logically establishes the truth of identity, by critical reflection and discourse. At the end of the stage of reflection all ignorance is removed, the aspirant is intellectually convinced of the identity of *Brahman* with *Atman*, and thus, moves towards the next stage, *viz.*, the practice of meditation.

Meditation

In this section we will briefly analyze the meaning and types of meditation. Besides, we will make an attempt to explain the practice of meditation, by aspirant's concentration on the *mahavakyas* and the eternal syllable '*Aum*'.

Meaning of Meditation: The final stage leading to the complete removal of ignorance and thus, to the direct realization of the self is meditation (*nididhyasana*). If an aspirant, having heard the teacher, is successful in his reflection and is intellectually convinced of his identity with *Brahman*, then he is ready to strive for the direct realization or *Brahmanubhava*. *Nididhyasana* is a mental activity consisting in withdrawing the mind from all other things and concentrating it on *Brahman*. It "strikes deep the Vedantic wisdom into our heart. It eradicates the innate confusion of the body with the soul." Meditation is not a concentration of oneself on *Brahman*, as an external of separate entity. It is an activity of the mind "in which the mind is turned completely inward, and is firmly, fixed on the inner self and its identity with *Brahman* till one's finitude and individuality is dissolved."

Thus, "meditation is the stream of ideas of the same kind as those of *Brahman*, the one without a second, to the exclusion of such ideas as those of body" and all other faculties related to the physical, psychical and intellectual dimensions of an individual. Hence, meditation involves a continuous and unbroken thought on *Brahman*, that flows like a line of flowing oil, and the exclusion of all thoughts. For Shankara, therefore, meditation consists in "remaining independent of everything as a result of the unassailable thought `I am verily *Brahman*' and [which] is productive of supreme bliss." Now that we have analyzed the meaning of meditation, we could talk about its different types in the next section.

Types of Meditation

Nididhyasana has two forms, *viz.*, *samprajnat-samapatti* and *asamprajnat-samapatti*. *Samprajnat-samapatti* is a form of meditation in which the aspirant experiences modifications of consciousness while meditating on the *mahavakyas*. In this type of meditation there are two stages. The initial stage is characterized by the knowledge of the modification of mental consciousness that originated while meditating on the Vedantic statements. The aspirant, therefore, is conscious of himself, the meditator and the witness of the modification that has taken place in the consciousness, and of the modification created by the meditation on the scriptural axiom at that particular moment.

The later stage of *samprajnat-samapatti* is free from all thoughts regarding the origin of the modification that is produced in the consciousness, as the result of meditation on the *mahavakya*. Since meditation is intense at this stage, the temporal and spatial marks of modifications are not available to the consciousness of the meditator. The aspirant is only aware of himself as the witness and the modifications produced by his meditation on the Vedantic aphorisms.

Asamprajnat-samapatti is a state of meditation, in which the consciousness of the aspirant practicing meditation is not characterized by any modifications. In it, there is no sense of duality as all modifications produced by the meditation, on the scriptural axiom, have ceased to exist. As there is no subject-object distinction in this meditation, the *Atman* becomes the subject and object of the meditation as the identity without any modifications is arrived at as the result of *asamprajnat-samapatti*.We have looked into the meaning and types of meditation in the last two sections. In the next section, we move on to consider the practice of meditation with the help of the four *mahavakyas* and the eternal syllable *'Aum'*.

Practice of Meditation: In the states of *sravana* and *manana* the aspirant studies the scriptures, analyses the meaning of the Vedantic aphorisms and discriminates the perishable body, mind, intellect and plurality of the world from the imperishable self. Having done this, the aspirant negates the perishable world and asserts the imperishable self as real, repeatedly suggesting to himself that "I am supreme *Brahman*".The latter task of assertion is done by the practice of meditation, which takes the aspirant to experience the fundamental identity of *Atman* and *Brahman*. The practice of meditation is facilitated by the aspirant's focus on the four *mahavakyas* and the eternal syllable `*Aum*, while meditating.

The aspirant's practice of meditation on the *mahavakyas* must begin with the *lakshana vakya* (statement of definition), *viz.*, '*Prajnanam Brahma*' ('Consciousness is *Brahman*'). This aphorism fixes the mind of the meditator on the thought that consciousness in the individual (*Atman*) and the consciousness underlying the entire universe (*Brahman*) are one and the same. In other words, meditating on this aphorism the aspirant recognizes existentially that the same consciousness is the substratum of the microcosm and the macrocosm. Thus, the meditation on the statement 'Consciousness if *Brahman*' envelops the depth of the aspirant with the experience that the consciousness ever remains the one homogeneous reality whether it be in the cosmos or in the individual. With the deep awareness of the oneness of the ultimate reality, the aspirant moves on to the meditation of the *upadeesha vakya* (statement of advice), *viz.*, '*Tat Tvam Asi* ('That art Thou'). It asserts that the one ultimate reality (That) is the pure self (Thou) that is the core of one's personality that lies beyond the five sheaths (*koshas*) of matter. In this manner the meditation on the second *mahavakya* makes the aspirant experience the oneness of the infinite *Brahman* and *Atman* within him. Having convinced that the supreme reality is nothing but his own self, the aspirant moves on to the meditation of the *abyhasa vakya* (statement of practice), *viz.*, '*Ayamatama Brahma*' ('This Self is *Brahman*).

The meditation on this practical formula makes the aspirant realize *Atman*, which activates him is the same *Brahman*, who vitalizes the entire universe. Thus, the aspirant discovers the identity between the self and the all-pervading *Brahman*. The conviction arrived at by the meditation on the third *mahavakya* makes the aspirant experience the truth of *anubhava vakya* (statement of experience), *viz.*, '*Aham Brahmasmi*' ('I am *Brahman*'). This statement is a pronouncement of the man of realization. He knows now experientially that he is the all-pervading *Brahman* and that all the duality is totally removed.

The 'I' referred to in the fourth *mahavakya* is different from the 'I' experienced in the waking state, dream state and the state of deep sleep. The 'I' of the *anubhava vakya* is the supreme self, identical with *Brahman*.

In this manner the practice of meditation facilitated by the *mahavakyas* lead to self-realization. Besides meditating on the *mahavakyas*, the aspirant must attempt to focus his attention on the eternal syllable '*Aum*', while practicing meditation. The meditation on '*Aum*' would remove all traces of ignorance and lead the aspirant to the gates of true knowledge. Manduukya Upanishad refers to the significance of the eternal syllable '*Aum*' as follows: "*Aum*, this syllable is all this. All that is past, the present and the future, all this is only the syllable '*Aum*'. And whatever else there is beyond the three fold time, that too is the only syllable '*Aum*'." This passage from the Manduukya Upanishad clearly states that '*Aum*' refers to both the manifested and the unmanifested *Brahman*.

The ultimate and unmanifested *Brahman*, the pure consciousness, though one in itself, by the activity of cosmic and individual *maya*, in the temporal order appears as Virat, *Hirayanagarbha* and *Ishvara* in the macrocosm; and *Vishva*, *Taijasa* and *Prajna* in the microcosm. Thus, '*Aum*' refers to the manifested *Brahman* conditioned by the three fold states both in the macrocosm and the microcosm. Therefore, the three sound elements of the syllable '*Aum*' corresponds to and signifies *Brahman* conditioned by the three bodies (*viz.*, the gross, subtle and causal bodies) and manifesting in the three states (*viz.*, the waking state, the dream-state and the deep sleep state).

So Shankara maintains that the sound '*A*' of the syllable '*Aum*' signifies *Virat* in the macrocosm and *Vishva* in the microcosm; the sound '*U*' of the syllable '*Aum*' represents *Hirayanagarbha* and *Taijasa*; and the sound '*M*' of the syllable '*Aum*' indicates the *Ishvara* and *Prajna*. In meditation the aspirant attempts to inculcate the corresponding relationship between different sound of the syllable '*Aum*' and the respective pairs of macrocosmic and microcosmic realms. In doing so, he is encouraged to dismiss consciously the differences between *Virat* and *Vishva*, *Hirayanagarbha* and *Taijasa*, and *Ishvara* and *Prajna*, and apprehend these pairs as identical. To quote Sureshvaracharya on this point: "The three forms, *Vishva*, *Taijasa* and *Prajna* must be contemplated as identical with *Virat*, *Suutratma* [*Hirayanagarbha*] and *Akshara* [*Ishvara*] respectively, so that the non-existence of the difference of those entities may be established." Thus the three individual forms of consciousness become identical with the three collective forms of consciousness, and so only the latter remains in place of the six.

When a seeker practices this identification of the macrocosmic and microcosmic states, by way of meditation he would experience, the three sounds of the syllable '*Aum*', *viz.*, '*A*', '*U*' and '*M*', as signifying the three phases of the one integral spirit, *i.e.*, *Brahman*. Thus, the whole universe would be seen as the ultimate reality, in relation to the waking consciousness, dream consciousness and deep-sleep consciousness conditioned by the gross body, the subtle body and causal body respectively. Even at this stage the aspirant of self-realization has two fold problems, *viz.*, a philosophical and a spiritual. The philosophical problem consists in understanding the ultimate *Brahman* as transcending the conditions in which it is seemingly embodied. The spiritual problem is to raise oneself from the limiting conditions of the gross body, the subtle body and the causal body and to realize one's identity with the ultimate *Brahman*.

In order to overcome these two problems and arrive at the identity experience, the aspirant must learn to move, both intellectually and spiritually, from the entanglements with the different bodies in different states of existence. This involves an ascending movement form the gross body in the waking level to the level of subtle body in the dream level and from the dream level to the deep-sleep level of the causal body. Constant meditation on the different sounds of the syllable '*Aum*', *viz.*, '*A*', '*U*' and '*M*', would bring about this upward movement towards the realization of *Brahman*. The first sound that is constitutive of '*Aum*', '*A*' represents the gross point of view of reality. Philosophically it refers to the naive realism and the pluralism of common sense. The aspirant must move from this pluralistic perspective of reality and from entanglement with the gross things that characterize his life in the waking state.

He must move towards the level of thought that is marked by subtle bodies that are characteristic of dream state. This state is represented by the sound *'U'* of *'Aum'*. Philosophically the thought belongs to this level is dynamic idealism, which sees the universe as the projection of the spirit, just as the dream state is the projection of the mind. Thus, one realizes at this state that the phenomenal reality is the manifestation of *Brahman*, by the instrumentality of *maya*. This state is as unreal as the waking state. When an aspirant attains maturity in this level, he must move onto the third level of the state of deep sleep that is characterized by causal body and ignorance. This state is represented by the final sound of *'Aum'*, *viz.*, the *'M'*. The philosophy of state of deep sleep is agnosticism, as in this state ignorance the seed of phenomenal existence is present in a striking manner. As an aspirant meditates on the *'M'* sound of *'Aum'*, he must break the agnosticism of the deep sleep state and experience the totality of the eternal syllable *'Aum'*, which is fundamentally and essentially the unmanifest *Brahman*.

Thus, meditating on the three sounds of the syllable *'Aum'*, the aspirant gradually recognizes the unreality of the waking state with the gross body, the dream-state with the subtle body and the deep sleep-state with the causal body. In this manner, he wakes up to the infinite reality of his spiritual essence in the experience of the eternal syllable *'Aum'*. Moving out of the three stages of ignorance, the aspirant achieves absolute illumination, which makes him experience his divine essence as the absolute *Brahman* in the 'stateless' eternity. Shankara in his work *Panchiikaranam* sums up the experience of the aspirant attaining self-realization by meditation on the eternal syllable *'Aum'* as follows:

> *Now* 'A' *the waking personality should be resolved into* 'U', *the dream personality, and the* 'U' *into* 'M', *i.e., the deep-sleep personality. Again the* 'M' *should be reduced into* 'Aum' *and the* 'Aum' *into "I". I am, the* Atman, *the Witness of all, the absolute, of the nature of pure*

> *consciousness; I am neither nescience nor even its effect but I am* Brahman *alone, Eternally Pure, Ever Enlightened, Eternally Free, and Existence Absolute. I am the Bliss Absolute, One without a second and the Innermost Consciousness. Sureshvaracharya commenting on this passage from Shankara's* Panchiikaranam, *says as follows:*

The waking personality of *Vishva* [-*Virat*], symbolized by 'A' must be resolved into 'U' (*i.e.*, the dream personality). The subtle radiant personality of the dream, the '*Taijasa*' [-*Hirayanagarbha*], symbolized by 'U' must be merged into 'M' (*i.e.*, the personality of the deep sleep). Again the *Prajña* [-*Ishvara*], that deep sleep consciousness symbolized by 'M' and which is the causal personality must be finally reduced to *Atman*, of the nature of Pure Consciousness. Such meditation on the eternal syllable '*Aum*', in which the aspirant dissolves and reduces the lower on the higher and finally everything in *Brahman*, the absolute self is traditionally known as *Lai Upasana* or *Ahangrah Upasana*. Along with the meditation on the *mahavakyas* the *Lai Upasana* is an excellent way of practicing meditation, which would lead to the dawn of knowledge, when the seeker experiences identity between the *Brahman* and the *Atman*.

Thus, by repeated exercise of meditation one moves to greater depth of absolute consciousness. This consciousness of identity with the absolute *Brahman* removes all the effects of ignorance. By focusing more on the inner self, by way of meditation the aspirant makes the inward journey, until he experiences his absolute identity with *Brahman*.

Release: The End of the Path

In the last two sections of this chapter, we considered the nature and methods of *Brahmajijnasa* and the various stages of Shankara's *jnana* path to self-realization. The preparations the aspirant makes in the physical, moral and intellectual levels, helps him focus more on the inner nature, wherein he finds the ultimate source of his existence. The end of this journey is the attainment of release (*Samadhi*). In this section, we make an attempt to elaborate the nature of the state of *Samadhi* and its goal, *viz.*, the emergence of the realized man (*Brahmajnani*)

Nature of Samadhi

In this section, we make an attempt to analyze the meaning of the state of release. Besides, we would briefly mention the different types of *Samadhi* and the obstacles the seeker has to overcome in order to attain this state, the final goal of human existence.

Meaning of Samadhi: The realization of one's absorption in or identity with *Brahman* is release (*Samadhi*). It is the true liberation and the ultimate end of the seeker. This state of self-realization is of the same nature of *Brahman*. Therefore, *Brahman* and *Samadhi* are identical. Liberation is nothing else, but becoming one with *Brahman*. In the liberated state the seeker knows that he is *Brahman*. As a result, all duality and multiplicity disappear. One knows now that all, including his self is *Brahman*. In *Samadhi*, nothing new is attained in the aspirant, for he only realizes what he is from all eternity. Speaking of the state of release Shankara says: "The complete forgetfulness of all thought, by first making it changeless and then identifying it with *Brahman* is called *Samadhi*, known also as knowledge." By this statement Shankara does not say that release is a state of unconsciousness. Even though all objective thoughts are absent in *Samadhi*, the pure consciousness is always there.

To deny the presence of consciousness at this state is an impossibility, as it is the very self of the person, who denies it. Since the very attempt to deny the presence of the consciousness at the state of release would affirm the reality of consciousness, this state can be rightly called knowledge. From what we have said, it is clear that *Samadhi* is different from the state of deep sleep. It is true that mental states or object-thoughts do not appear both in *Samadhi* and in deep sleep. But the fundamental difference between the two consists in that in *Samadhi* the mental state exists having taken the form of *Brahman*, while in deep sleep the mental state is totally absent, as it has merged with ignorance which alone remains. Thus, the state of release is different from and essentially superior to the state of deep sleep. According to Shankara, the realization of *Samadhi* takes place in three stages of consciousness.

The first stage is *asmbhava-bhavana*. It consists in the removal of thought of non-existence of *Brahman*, when one hears that '*Brahman*, as undivided consciousness, exists'. The second stage is *drishyamarjnana*. Here, the discriminative capacity is more advanced and one is able to penetrate the appearance to get into the essence of reality. Now, one is able to remove the *avidhya* that everything is material and become conscious of the immutable *Brahman*. At these stages, the knowledge is only indirect, mediate and based on subject-object distinction. The third stage is that of identity-consciousness. At this stage, the seeker experiences the deepest core of his being that *Atman* is identical with *Brahman*, the ultimate source behind the universe. In the process, all ignorance about the illusory nature of the phenomenal reality is removed. This final stage gives us direct knowledge of *Brahman*. Commenting on these three stages of consciousness Mahendranath Sircar says:

> *The first stage marks out the origin and continuity of* vritti, *the second, its final disappearance, the third is the stage of knowledge. Between the second stage and the expression of* Atman *in the third, if we can speak in such a way, there is no sequence of time. They are simultaneous.* Atman *is known only by implication as one invariably associated with the denial of illusory forms of* avidhya *and of* avidhya *itself. According to Vedantins there is a difference between the perception of* Brahman *in* Brahmanubhava *and the perception of concrete facts in the phenomenal existence.*
>
> *In the perception of concrete things, the mind goes out and takes on itself the determination of the object. Here, the perceiver become aware of the existence of the object, because of the fact that mind takes the form of the cbject, and thereby removes the perceiver's concrete ignorance about the object. If we take the example of the pot as the*

object of consciousness, it is known because the mind goes out through the senses and takes on itself the form of the pot. This modification of the mind removes the ignorance about the pot. As a result, consciousness expresses the object. This process is technically known as falavapya. *But in the perception of* Brahman *in* Samadhi *is very different.*

In it, the mind does not take any concrete form, as Brahman *has no form.* Brahman-*experience in* Samadhi *is* vrittivapya. *So the mind does not go out to experience* Brahman, *but rather it is transformed into* Brahman, *putting an end to all forms of ignorance. Since we have clarified the meaning of the state of release, we could analyze its different types in the next section.*

Types of Samadhi: The later Advaitic writers speak of two types of *Samadhi* depending on the nature and intensity of concentration. They are determinate absorption (*Savikalpa Samadhi*) and indeterminate absorption (*Nirvikalpa Samadhi*). In determinate absorption the aspirant abides in *Brahman*, but still retains his 'I-consciousness'. Here, there exists the distinction between 'the knower', 'the known' and 'the knowledge'. "Absorption attended with self-consciousness (*Savikalpa Samadhi*) is that in which the mental state taking the form of *Brahman*, the One without a second, rests on It, but without merging of the distinction of the knower, the knowledge and the object of knowledge."

Thus, determinate absorption is a state of *Brahman*-experience, in which the mind assumes the form of *Brahman* and rests on it with the distinction between subject and object still persisting. In other words, though the mind of seeker abides in *Brahman*, the contemplative faculty has not merged into the final witness, *i.e.*, the absolute *Atman* and as a result the three fold division of meditation, meditated and meditator, still persists in the seeker's mind in *Savikalpa Samadhi*. Thus, in determinate absorption the knowledge of absolute *Brahman* manifests itself, in spite of the fact that the consciousness is relative. The later Advaitins compare this state to one who possesses knowledge of both clay elephant and the clay that permeates the elephant. In this state, therefore, both the phenomenon and the noumenon (absolute substratum) are present before the seeker's mind, and the absolute reality peeps through the vestige of name and form.

In other words, in the *Savikalpa Samadhi* the seeker is already possessed by the truth, but still unable to realize it entirely. *Nirvikalpa Samadhi* (indeterminate absorption) is the total absorption into *Brahman*. When an aspirant, having practiced *Savikalpa Samadhi* for a long period of time, loses the sense of duality of the subject and the object altogether and becomes one with *Brahman*, he is said to have attained the state of *Nirvikalpa Samadhi*. It is an "absorption without self consciousness [and it consists in] the total mergence in *Brahman*, the One without a second, of the mental state which has assumed Its form, the distinction of knower, knowledge, and the object of knowledge being in this case obliterated." In this state, the contemplative faculty itself is totally dissolved into *Brahman*. As a result all distinctions between meditation, meditated and the meditator totally disappears and there remains only the eternal self.

"Just as when salt has been dissolved in water it is no longer perceived separately, and the water alone remains, similarly the mental state that has assumed the form of *Brahman*, the One without a second, is no longer perceived, and only the self remains." Thus, in this state there is no 'I-consciousness' and no subject-object duality, as the meditative faculty is totally merged in *Brahman*. The *Panchiikarana-Varttikam* of Sureshvaracharya states about the *Nirvikalpa Samadhi* as follows:

> *When the contemplative mind is merged into* Atman, *the Pure Consciousness, then it should not be disturbed. One should remain as that Infinite Consciousness like the full and motionless motion. Thus attaining perfect absorption through constant practice, an aspirant endowed with faith and devotion and having overcome the senses and anger perceives (realizes) the* Atman, *One without a second. Thus, in the state of* Nirvikapa Samadhi, Brahman, *the ultimate self shines by its own*

radiance. The aspirant having totally dissolved everything in Brahman *experiences his absolute and total identity with* Brahman.

In order to attain the state of spiritual absorption in *Nirvikalpa Sammadhi*, thc aspirant must attempt to overcome a number of obstacles. To quote Shankara: "While practicing [*Nirvikalpa*] *Samadhi* there appears unavoidably many obstacles, such as lack of inquiry, idleness, desire for sense pleasure, dullness, distraction, tasting of joy and sense of blankness. One desiring the knowledge of *Brahman* should slowly get rid of such innumerable obstacles." Advaitins speak of four obstacles that an aspirant must work on in order that the state of *Nirvikalpa Samadhi* is achieved. They are torpidity (*laya*), distraction (*vikshepa*), attachment (*kashaya*) and enjoyment (rasavada). Torpidity consists in the lapse of mental state into sleep or a sense of blankness because of one's failure to rest on the absolute. It is the result of laziness and fatigue in the aspirant. Distraction is the resting of one's mental state on things other than the absolute *Brahman*.

It happens due to the lack of concentration on the part of the aspirant. When he fails to focus his entire attention to rest on *Brahman*, he is distracted. Attachment is the failure of one's mind to rest on the absolute *Brahman* due to the re-experiencing of the impressions (*vasanas*) of sense pleasures experienced earlier. Even when there is no torpidity or distraction and everything seems to be calm, attachment can torment an aspirant, and as a result he will be prevented from fixing his mind on *Brahman*. *Enjoyment* consists in the mind getting caught up in the state of bliss experienced by the seeker in the *Savikalpa Samadhi*. As a result the aspirant is not able to rest the mind on *Brahman*.

It can also happen when the aspirant enjoying the *Savikalpa Samadhi*, having no inner strength to give up the bliss and to give himself to the practice of *Nirvikalpa Samadhi* with full mind and heart. In both the cases this feeling of enjoyment greatly hinders the spiritual progress of the aspirant, as he is unable to rest totally and fully on the absolute *Brahman* and thereby attain the *Nirvikalpa Samadhi*. When the mind of the aspirant is free from all these four obstacles, and rests fully in the absolute consciousness, just like a flame of a lamp sheltered from the wind, he would be moving towards the state of *Nirvikalpa Samadhi*. The aspirant is called to do the following in order that he can remove these obstacles. When his mind is torpid, he must arouse and awaken it. When he is distracted, by perseverance and renunciation bring his mind to a state of calmness.

When the mind is attached to past experiences of pleasures, become aware of such impressions and direct the attention to the absolute *Brahman*. When the mind continues to linger on the spiritual enjoyment of the bliss of *Savikalpa Samadhi*, get unattached through discrimination of the lower from the higher. When the aspirant experiences a state of mental equipoise, he must continue to remain in it, never attempting to disturb himself from that state. When an aspirant takes all these measures to overcome the obstacles, he would experience the absolute state of *Nirvikalpa Samadhi*, where he experiences his self as totally identical with the absolute *Brahman*.

If we accept the distinction of the two different states in *Samadhi*, as proposed by the later Advaitins, then, without any doubt the latter state, *Nirvikalpa Samadhi* is the same as self-realization (*Brahmanubhava*). Here, the absolute identity between the seeker and *Brahman* is realized. Such a realized seeker becomes a *Brahmajnani*. The emergence of *Brahmajnani* out of the aspirant of self-realization is the goal of the state of *Samadhi*. In the next section we turn our attention to this topic.

Goal of Release—Brahmajnani: When the state of release is attained, the seeker becomes a *Brahmajnani*. Thus, the goal of the state of *Samadhi* is the birth of the *Brahmajnani*. In this section, we attempt to analyze the nature of *Brahmajnani* and the two possible stages of his existence as the *Brahmajnani*, *viz.*, *Videhamukti* and *Jivanmukti*. Besides, this section would elaborate the nature and characteristic of *Jivanmukta*.

Nature and Stages of Brahmajnani's Existence: Here, we highlight the nature of the existence of *Brahmajnani* in a general way. Besides, we would look into the two possible stages of the existence of *Brahmajnani, i.e., Videhamukti* and *Jivanmukti*.

Nature of Brahmajnai's Existence: *Brahmajnani* is one who possesses the *jnana* of *Brahman*. He is in the state of transcendental consciousness. He would be purely unconscious of thc cmpirical order, as it would not affect him in any way. The phenomenal world, its variety of realities and their meanings no longer obstruct the vision of *Brahmajnani*. His state of existence cannot be described in positive terms, as it surpasses any type of description. He is enlightened, free, fully unaffected by pains and gains of the phenomenal existence. In his innermost essence, he knows that he is that eternal consciousness, ultimate truth and bliss. Thus, *Brahmajnani* is a man-of-realization, who has realized his supreme self. He has given up all his thoughts, desires and their impressions (*vasanas*), and has become his real self. Neither the material faculties of body, mind and intellect bind the *Brahmajnai*, nor the three stages of existence, *viz.*, the waking stage, the dream stage and the state of deep sleep limit his life, as they used to before the attainment of self-realization. *Brahmajnani* has overcome all limitations of all these stages and the bodies associated with them, *viz.*, the gross body, the subtle body and the causal body, respectively, and has become one with *Brahman*. Having crossed all these stages and reached the pinnacle of perfection, the *Brahmajnani* knows the unreality of all these stages that marked his state of ignorance.

He knows, now, that he is the eternal *Brahman*, pure consciousness and substratum of every phenomenal existence and stages of existence. *Brahmajnani* becomes the true witness of the microcosmic and macrocosmic stages of stages of existence, *viz.*, *Vishva*, *Taijasa*, *Prajna*, *Virat*, *Hirayanagarbha* and *Ishvara*. As a being-in-the-world, the man-of-realization experiences all these conditioned stages, but he remains unaffected by them as he knows that he is *Brahman*, and there is no more reality to be known."For him {*Brahmajnani*] who sees the all-pervading *Atman*, of the supreme peace and bliss and the sole reality there remains nothing more to be attained and nothing more to be known."

There is nothing more to be attained for the *Brahmajnani*, as in knowing the bliss of *Brahman*, there is nothing achievable left behind. In the same way, there is nothing more to be known for the *Brahmajnani*, as in knowing him to be *Brahman* nothing more remains to be known. Thus, "a wise one [*Brahmajnani*] attains the acme of life having nothing more to be achieved, and thus becomes eternally free although still living. With the whole of his mind and heart thoroughly filled with *Atman*, he does not perceive this world." *Brahmajnai*, therefore, is one who has allowed all the pluralistic manifestations of the microcosm and the macrocosm to merge into his eternal self, which is identical with *Brahman*. Having looked into the nature of the *Brahmajnani*, we could speak of the two stages of his existence in the following section.

Stages of Brahmajnani's Existence: We can speak of two stages of the existence of *Brahmajnai*. When *Brahmajnani* reaches the transcendental peak of existence, he is called *Videhamukta* and his state is called *Videhamukti*. For the *Videhamukta*, the empirical world is not more a reality, as he has awakened to the new vision of existence in which every form of illusion is removed. *Videhamukta* is said to have attained *Videhakaivalya*, which involves freedom of being alone undisturbed, denial of the body and forsaking of future life. In this state, the *karamic* seeds that leads to future births have been destroyed, by the clear vision of identity. The effects of past *karma* have been obliterated.

As a result, there is no reason for the individual to continue living in the phenomenal world. *Videhamukta* passes into a calm existence, having lost to the empirical world and shedding his artificial personality that is characteristic of the *jiva*. Such a state is identical with the dawn of knowledge about *Brahman*. Thus, *Videhamukta* reaches the state of complete liberation (*kaivalya*) from which there is no return, as *karmic* forces are totally obliterated." Shankara is of the opinion that *Samadhi* or *Brahmanubhava* is possible even when one is alive. He is of the view that everyone can attain this state of identity with *Brahman* and thereby become a *Brahmajnai*, here in this life. All that is required of an aspirant is to the practice the various steps of *Brahmajijnasa* and work on removing the ignorance, the cause of duality. In Bhagwad Gita Bhashya, Shankara says: "A *yogi* (in the Advaitic sense) attains *Brahmanirvana* (same as *Brahmanubhava*), the bliss of being *Brahman* or liberation, by being *Brahman* here itself, that is while he is alive." Again commenting on the Upanishadic statement "by being *Brahman* one attains *Brahman*", Shankara says that *Brahmanubhava* is possible while one is alive and there is no need for one to wait until death for its attainment.

When ignorance is fully destroyed, the real nature of the self is revealed. Such a *Brahmajnani*, who enjoys liberated state in the present existence, is called *Jivanmukta*. The state of liberation he enjoys is called *Jivanmukti*. In fact, death does not change the essential condition of the *Jivanmukta*, but only brings to a close the effects of the accumulate action which is still bearing fruit (*prarabdha karma*). In other words, death puts an end to the present life, which is the effect of *prarabdha karma*.

Thus, the fundamental difference between *Videhamukta* and *Jivanmukta* is that while the former is fully free from *prarabdha karma*, the latter is still under its sway. This is very clearly stated in the following quotation from *Vedantasara*:

> *Such a man's [*Jivanmukta's*] soul remains as the illuminer [the witness of everything] of the mental states and the consciousness reflected in them experiencing solely for the maintenance of his body, happiness, and misery, the result of past actions that have already begun to bear fruit (*prarabdha*) and have been either brought on by his own will or by that of another or against his will. After exhaustion of the* prarabdha *work [through enjoyment and suffering], his vital force is absorbed in the Supreme* Brahman, *the Inward Bliss [and he becomes* Videhamukta*]; and ignorance with its effects and their impressions is also destroyed.*
>
> *Then he is identified with the Absolute* Brahman, *the Supreme Isolation, the embodiment of Bliss, in which there is not even the appearance of duality. Thus,* Videhamukti *and* Jivanmukti *are not essentially*

different stages of Brahmajnai, *but it is basically same and the only difference between the two is the non-operation and operation of the* prarabdha karma *respectively. Having clarified the nature and stages of existence of* Brahmajnani, *we could move on, in the next section, to analyze in detail the nature and characteristics of* Jivanmukta, *i.e., the life of a realized man living this earthly existence.*

Jivanmukta

We have established that *Brahmanubhava* is possible for everyone, even in this life, there arises the question of the possibility of the behaviour of a *Jivanmukta*. Since there is no duality in this transcendental existence, is it possible for the liberated man to live in this world of duality? Even if it is possible for him to live in this world, what is his nature? What are some of his basic characteristics? How is he different from an ordinary unrealized person? In this section, we will attempt to answer these basic questions regarding *Jivanmukta*, by clarifying his nature and elaborating his characteristics.

Nature of Jivanmukta: The behaviour of the liberated man can be distinguished from that of others on the basis of the absence of ignorance and its effects. The fully realized soul does not possess any trace of ignorance and sees everything in *Brahman*. Thus, the behaviour of the *Jivanmukta* is characterized by oneness, while an ignorant person's behaviour is based on experience of differences. Though *Jivanmukta* lives in the world of duality none of the pairs of opposites disturb him as he sees all thing in terms of oneness with *Brahman*.

Nothing affects him, since he sees everything in himself. An ignorant person considers others and different from himself and shapes his relationship with them accordingly, whereas the liberated man does not see others as different from himself and shapes his relationships with others in terms of oneness. Thus, the liberated man is "one who by the knowledge of the absolute *Brahman*, his own self, has dispelled the ignorance regarding It, and has realized It, and who owing to the distinction of ignorance and its effects such as accumulated past actions, doubts, errors, etc., is free from all bondage and is established in *Brahman*." Since *Jivanmukta* sees everything in relation to his own self, the absolute *Brahman*, nothing can bring any change in his own self.

By his very nature *Jivanmukta* is fearless. He cannot be afraid of anything. For Shankara says, "Fear is caused by the second entity or by things conceived to exist as different from the self. And when this notion of a second entity is eliminated by the realization of oneness, there will be no source of fear." An ignorant person sees everything in terms of differences, and so there is every reason that there is fear in the person caught up by the phenomenal existence. But for the one who has realized *Brahman*, the absolute and indestructible self, there is nothing to be afraid of, as he is the one and the absolute.

Jivanmukta transcends scriptures, ethical imperatives and social conventions. As an aspirant, he while working towards this ultimate realization eradicated all his passions, prejudices and attachments, and concentrated on *Brahman* alone. Thus, when he has attained the identity with *Brahman*, he is free from all faults and never makes false steps or sets a bad example. The fundamental difference between an aspirant for liberation and a *Jivanmukta* is that the scriptural injunctions and moral practices bind the former, while the latter is neither a slave of laws, nor wantonly violates them. He is beyond laws and morality.

But great ethical virtues such as humility, unselfishness, purity, kindness and fellow feeling which prior to the attaining knowledge, he assiduously practiced for the purification of the mind, now, adorn him like jewels. He does not seek them or need them, but they cling to him."After realization, humility and other attributes which are steps to the attainment of knowledge, and also such virtues as non-injury, etc., persist like so many ornaments. [Again] such qualities as non-violence, etc., come spontaneously to a man who has got self-knowledge [*Jivanmukta*]. They have not to be sought after."

The liberated man alone knows the true nature and meaning of freedom. He is free from all types of bondage imposed on man. He is the all-embracing self, and is absolutely free from the cares and worries of life. Shankara portrays the freedom enjoyed by the liberated man as follows:

> *Sometimes he [*Jivanmukta*] appears to be a fool, sometimes a wise man. Sometimes he seems splendid as a king sometimes feeble minded. Sometimes he draws men to him, as a python attacks its prey. Sometimes people honour him greatly sometimes they insult him. Some times ignore him. That is how the illumined soul lives, always absorbed in the highest bliss. He has no riches, yet he is always contented. He is helpless and yet of mighty power. He enjoys nothing, yet he is continuously rejoicing. He has no equal, yet not bound by action. He reaps fruits of the past action, but does not identify himself with it. He appears to be an individual, yet he is present in all things, everywhere. The knower of* Brahman, *who lives in freedom from body-consciousness, is never touched by pleasure or pain, good or evil. Thus, whatever may be the state or condition he is in,* Jivanmukta *is free to conduct himself according to that condition.*
>
> *He is, in himself, the absolute and lacks nothing, and so is disturbed by nothing whatsoever. So* Jivanmukta *enjoys a freedom which is not found in a man of ignorance.*

The *Jivanmukta* is without any desire and free from sorrow and grief. Desire arises, when the object of one's wish is not attained. Sorrow and grief arise when the object of one's desire is no longer with him. A person is sad because he has lost something, which was dear to him. In fact, desire, grief and sorrow are based on the experience of differences. *Brahman* is the absolute self and it lacks nothing. In attaining the self everything else is attained, for *Brahman* is the ultimate source of everything.

Thus, absence of desire, grief and sorrow, in case of the *Jivanmukta* is not due to suppression of desires but because of his realization of *Brahman*, after reaching which, there remains nothing to be desired. To quote Shankara "The things perceived by the senses cause him [*Jivanmukta*] neither grief nor pleasure. He is not attached to them. Neither does he shun them. He lives desireless amidst the objects of desire. The *Atman* is his eternal satisfaction. He sees the *Atman* in all things." Therefore, there is nothing left to desire and there is nothing that can bring him sorrow.

The *Jivanmukta* lives in this bodily state as long as there lasts the accumulated effects of the past action that have begun to bear fruit (*prarabdha karma*). Until that time *Jivanmukta* might engage himself in working for the welfare of others. As a possessor of the body, which is the result of earlier *karma*, he experiences that which is characteristic of material forms like hunger, thirst, illness, and old age. But, these never overwhelm him, for he knows the truth of their passing nature and of his nature as the absolute *Brahman*. *Vedantasara* comments on this point as follows:

> *Such a liberated man, while not in* Samadhi, *sees actions not opposed to knowledge taking place under the momentum of past impressions—actions that have already begun to bear fruit which he experiences through the physical body composed of flesh, blood and other things; through the sense organs affected by blindness, weakness, incapacity, etc. and through his mind subject to hunger, thirst, grief, delusion, etc.—yet he does not consider them as real, for he has already known their nothingness.*
>
> *As a man who is conscious of that magical performance is being given, even though he sees it, does not consider it as real. Therefore, the liberated man, the true knower of the self, is one, who "sees nothing in the waking state, even as in dreamless sleep; who though beholding duality, does not really behold it, since he beholds only the absolute; who though engaged in work is inactive."*
>
> *In this section, so far, we have elaborated the behaviour that is typical of the nature of the* Jivanmukta. *In the next section, we could proceed to spell out some of the significant characteristics of the person who has attained self-realization.*

Characteristics of Jivanmukta: The man of realization is one with the supreme *Brahman*. His individuality is fully developed, as it has fully merged with the all-pervading absolute consciousness. With the self-realization, the *Jiinvanmukta* acquires a number of qualities and attributes. Herein we could mention some of the basic characteristics of man-of-realization.

Independence is one of the marked traits of a *Jivanmukta*. This attribute emerges from the fact that the realized man is not dependent on his body and its perceptions, his mind and its affection, and his intellect and its thoughts. As long as a person is attached to these physical faculties, he is affected by physical conditions, swayed by emotions and coloured by ideas. Thus, the identification of oneself with the body, the mind and the intellect is the reason for one's dependence on the world.

But a *Jivanmukta* has detached himself from these faculties of the body and identified himself with the supreme self. Therefore, the world cannot touch him. As he is free from the physical, mental and intellectual affections and modifications, *Jivanmukta* remains pure and uncontaminated by the world.

The Vedantins use the illustration of a coconut to clarify the independence that marks the life of a self-realized person. When a coconut is raw, the kernel of the coconut sticks to the shell, as it cannot exist apart from the shell. Breaking the shell would bring about the breaking of the kernel as well. But when the coconut dries up completely, the kernel gets separated from the shell, and it shakes within the shell. If one breaks the shell now the kernel does not break and it remains unhurt, as it is free from the attachment to the shell.

This is the same with the self. So long as one identifies his self with the body-mind-intellect system his life is affected by the world. His life can be compared to a raw coconut that sticks to the shell. But *Jivanmukta* is like the kernel that is totally independent of the shell, remaining ever unaffected by the affections of the world, enjoying a bliss of *Brahman*, which his true self.

Jivanmukta is a man of universal love. In him there is no trace of selfishness. His love is truly other-centred. The love people have for each other in the phenomenal living is often not true love. It is an uneven or preferential love, which is basically a love that involves selfishness. If we look into the core of this preferential love, we find nothing else but self-love.

For example when a man loves his sun dearly, his love is fundamentally directed to his own well being. True love is not preferential, but universal and it is same for one and all. The interest, attention and identification involved in the love of the *Jivanmukta* is for all and yet to everyone in particular. This universal love is not limited to or bound by caste, creed, colour, community or country. It is an unfixed and all-pervading love, yet reaches to every individual, without any fixation and attachments on the part of the realized man.

Thus, the self-realized man is an embodiment of divine love that is free from all worldly divisions and separations. Another striking quality that we find in *Jivanmukta* is the sense of objectivity. It means to maintain an impersonal attitude towards every situation in life. It implies being distanced, unattached and uninvolved. Being objective does not mean lack of concern for people around. But it means a distanced and an uninvolved concern.

The realized man is truly concerned about the welfare of everyone around him, but in judgment about people and situations he is objective and uninvolved. He is concerned about the world but not entangled with it. Nothing disturbs the inner serenity of his being one with absolute self.

Being one with the universal self, the *Jivanmukta* watches perceptions, emotions and thoughts pass by without any attachment to them. He treats his body, mind and intellect as things other than his true self. Thus, he becomes an eternal witness (*sakshi*) to all happenings in the world. The *Jivanmukta* is like the sun. He does everything in the world, but he is not entangled with anything in the world. The sun brings about life and luster, everywhere. It makes the plants grow, water flow, wind blow, evaporates water and brings rain. Though the sun does all these activities in the world, it is not at all involved in the world. Its presence is objective and distanced to the world in which it acts.

It makes things happen in the world, without getting involved. Thus, the sun is an uninvolved witness. The man-of-realization, like the sun, makes everything right in the world, eradicates sorrows and miseries of those who come to him and brings light of joy to all, yet living in the totally unruffled and undisturbed state of mind. Resting in his true self, the realized man looks at life, events, happenings and people without any prejudice and bias. He realizes the objective truth that all is *Brahman* and rests secure in this truth.

Resting secure in the truth that 'all is *Brahman*' the *Jivanmukta* experiences total fulfilment in his life. There is nothing that remains to be satisfied for a realized man. He is totally filled with *Brahman* that he aspires for nothing in this world. Neither physical pleasure, nor emotional joy and intellectual delight affect the *Jivanmukta*. This fulfilment of the man-of-realization comes from his knowledge that he is everywhere and everything, and that the world cannot give him any joy. The *Jivanmukta* is absolutely full (*paripuurna*) and lacks nothing. Because he experiences fulfilment in *Brahman*, *i.e.*, in the bliss of the self, neither the loss nor any gain would in any way affect the realized man. So, in him there is no stress or strain, no tension of any kind, no grief, sorrow or guilt. He is totally complete and fulfilled. He experiences a sense of self-sufficiency, which leaves him in a state of permanent joy within, that nothing of his experiences in the world affects him.

The realized man experiences a sense of fulfilment in *Brahman*. Two characteristics, *viz.*, cheerfulness and dynamism, mark this self-fulfilled existence. The cheerfulness that is manifested by the *Jivanmukta* is an outward expression of the inner fulfilment and bliss he experiences by his oneness with *Brahman*. It helps the realized man to be in the state of equipoise and joy in spite of many troubles he may experience in his life. Though cheerfulness is a distinguishing mark of the realized man, it is not often externally manifested.

Even if unexpressed, this cheerfulness remains a permanent feature in the realized man. The cheerful *Jivanmukta* does not remain just living in his own joy and bliss, but moves on in a dynamic way to work for the removal of evil and miseries that people experience in the world. The spiritual energy generated by *Brahman*-realization remains a dynamic force within the realized man, that he wishes and works for giving the absolute experience he had to all, as long as he lives in this world.

The quality of dynamism expresses itself in the physical, mental and intellectual levels of the *Jivanmukta*, as he attempts to reach out with the mission of giving the *Brahman*-experience to all. All we have said regarding the nature and characteristics of a *Jivanmukta* are only approximations. Just as *Brahman* and *Brahmanubhava* are incomprehensible and indescribable, so too the nature and characteristics of *Jivanmukta* are not describable. All we can say about the *Jivanmukta* and his behaviour is only from phenomenal point of view.

All we have done in trying to describe the nature and qualities of the *Jivanmukta* is to negate qualities like fear, desire, duality, differences, dependence, selfishness and similar attributes that are characteristic of those live under the sway of ignorance. In other words, we have only said what the *Jivanmukta* is not rather than what he is. He, like *Brahman*, is indescribable.

Therefore, the so-called characteristics mentioned are only one possible way of taking about the *Jivanmukta*, *i.e.*, from the phenomenal point of view. From the absolute (*paramartha*) point of view, *Jivanmukta* is *Brahman* and is of the nature of the unknowable and indescribable *Brahman*.

8

Concept of Maya in Hindu Religion

In Advaita Vedanta philosophy, maya is the limited, purely physical and mental reality in which our everyday consciousness has become entangled. Maya is believed to be an illusion, a veiling of the true, unitary Self—the Cosmic Spirit also known as Brahman. The concept of Maya was expounded in the Hindu scriptures known as the Upanishads. Many philosophies or religions seek to "pierce the veil" in order to glimpse the transcendent truth, from which the illusion of a physical reality springs, drawing from the idea that first came to life in the Hindu stream of Vedanta.

Maya is neither true nor untrue. Since Brahman is the only truth, Maya cannot be true. Since Maya causes the material world to be seen, it cannot be untrue. Hence, Maya is described as indescribable. She has two principle functions—one is to cover up Brahman and hide it from our mind. The other is to present the material world instead of Brahman. She is destructible. Consider an illusion of a rope being confused as a snake in the darkness. Just as this illusion gets destroyed when true knowledge of the rope is perceived, similarly, Maya gets destroyed for a person when they perceive Brahman with transendental knowledge.

A metaphor is also given—when the reflection of Brahman falls on Maya, Brahman appears as God (the Supreme Lord). In the pragmatic level, where the world is regarded as true, Maya becomes the divine magical power of the Supreme Lord, to create and rule the world. Maya is God's pious servant. God is not bound by Maya, just as a magician is not illusioned by their own magic. Hence, God is Bliss. However, unenlightened jiva are the servants of Maya, hence they are in misery.

In Hinduism, Maya must be seen through in order to achieve moksha (liberation of the soul from the cycle samsara) — ahamkar (ego-consciousness) and karma are seen as part of the binding forces of Maya. Maya is seen as the phenomenal universe, a lesser reality-lens superimposed on the one Brahman that leads us to think of the phenomenal cosmos as real.

By Sri Shankaracharya:

1. The Supreme Self (or Ultimate Reality) who is Pure Consciousness perceived Himself by Selfhood (*i.e.* Existence with "I"-Consciousness). He became endowed with the name "I". From that arose the basis of difference.
2. He exists verily in two parts, on account of which, the two could become husband and wife. Therefore, this space is ever filled up completely by the woman (or the feminine principle) surely.
3. And He, this Supreme Self thought (or reflected). Thence, human beings were born. Thus say the Upanishads through the statement of sage Yajnavalkya to his wife.
4. From the experience of bliss for a long time, there arose in the Supreme Self a certain state like deep sleep. From that (state) Maya (or the illusive power of the Supreme Self) was born just as a dream arises in sleep.
5. This Maya is without the characteristics of (or different from) Reality or unreality, without beginning and dependent on the Reality that is the Supreme Self. She, who is of the form of the Three Guna (qualities or energies of Nature) brings forth the Universe with movable and immovable (objects).
6. As for Maya, it is invisible (or not experienced by the senses). How can it produce a thing that is visible (or experienced by the senses)? How is a visible piece of cloth produced here by threads of invisible nature?

7. As there is the emission of the generative fluid on to a good garment on account of the experience of copulation in a dream, the pollution of the garment is seen as real on waking while the copulation was not true, the man in the dream was real (while) the woman was unreal and the union of the two was false (but), the emission of the generative fluid was real, so does it occur even in the matter in hand.
8. Thus Maya is invisible (or beyond sense-perception). (But) this universe which is its effect, is visible (or perceived by the senses). This would be Maya which, on its part, becomes the producer of joy by its own destruction.
9. Like night (or darkness) Maya is extremely insurmountable (or extremely difficult to be understood). Its nature is not perceived here. Even as it is being observed carefully (or being investigated) by sages, it vanishes like lightning.
10. Maya (the illusive power) is what is obtained in Brahman (or the Ultimate Reality). Avidya (or nescience or spiritual ignorance) is said to be dependent on Jiva (the individual soul or individualised consciousness). Mind is the knot which joins Consciousness and matter.
11. Space enclosed by a pot, or a jar or a hut or a wall has their several appellations (*e.g.*, pot space, jar space etc.). Like that, Consciousness (or the Self) covered here by Avidya (or nescience) is spoken of as jiva (the individual soul).
12. Objection: How indeed could ignorance become a covering (or an obscure factor) for Brahman (or the Supreme Spirit) who is Pure Consciousness, as if the darkness arising from the night (could become a concealing factor) for the sun which is self-luminous?
13. As the sun is hidden by clouds produced by the solar rays but surely, the character of the day is not hidden by those modified dense collection of clouds, so the Self, though pure, (or undefiled) is veiled for a long time by

ignorance. But its power of Consciousness in living beings, which is established in this world, is not veiled.

Understanding Maya Through Bhagavad Gita Verses

Bhagavad Gita, Ch.14, Verse 3. "My womb is the great Nature (Prakriti or MAYA). In that I place the germ (embryo of life). Thence is the birth of all beings".

Bhagavad Gita, Ch. 14, Verse 4 "Whatever forms are born, O Arjuna, in any womb whatsoever, the great Brahma (Nature) is their womb and I am the seed-giving father."

Explanation: Prakriti (Nature), made up of the three qualities (Sattwa, Rajas and Tamas), is the material cause of all beings.

In the great Prakriti, I place the seed for the birth of Brahma (the creator, also known as Hirayanagarbha, or Ishwar, or the conditioned Brahman); and the seed gives birth to all beings. The birth of Brahma (the creator) gives rise to the birth of beings.

The primordial Nature (prakriti) gives birth to Brahma, who creates all beings.

(I am the father; the primordial Nature is the mother).

Bhagavad Gita, Ch.13, verse 26. "Wherever a being is born, whether unmoving or moving, know thou Arjuna, that it is from the union between the field and the knower of the field". (Purusha is the knower of the field; Prakriti is the field; Shiva is another name for the knower of the field and Shakti is the field; Spirit is another name for the knower of the field and Matter (Prakriti) is the field).

Bhagavad Gita, Ch. 7, Verse 4. "I am endowed with two Shaktis, namely the superior and the inferior natures; the field and its knower (spirit is the knower of the field; matter is the field.) I unite these two".

Bhagavad Gita Ch.7, Verse 6. "Know these two-my higher and lower natures-as the womb of all beings. Therefore, I am the source and dissolution of the whole universe".

Bhagavad Gita, Ch.13, Verse 29. "He sees, who sees that all actions are performed by nature alone, and that the Self is action less".

(The Self is the silent witness).

Bhagavad Gita, Ch.9, Verse 17. "I am the father of this world, the mother, the dispenser of the fruits of actions and the grandfather; the one thing to be known, the purifier, the sacred monosyllable (AUM), and also the Rig, the Sama and the Yajur Vedas".

Maya in Hindu Mythology

Maya is also the name of an Asura, who was the father-in-law of the Lord of Lanka, Ravana and the father of Mandodari. He is the archnemesis of Vishwakarma, the celestial architect of the Gods. His knowledge and skills are compatible with Vishwakarma. When Lanka was destroyed by Hanuman, it was the King of Demons, Maya, who had re-installed the beauty of that Island Kingdom.

Maya as the Goddess

In Hinduism, Maya is also seen as a form of Lakshmi, a Divine Goddess. Her most famous explication is seen in the Devi Mahatmyam, where she is known as Mahamaya.

Essentially, Mahamaya (great Maya) both blinds us in delusion (moha) and has the power to free us from it. Maya, superimposed on Brahman, the one divine ground and essence of monist Hinduism, is envisioned as one with Lakshmi, Durga, etc. A great modern (19th century) Hindu sage who often spoke of Maya as being the same as the Shakti principle of Hinduism was Shri Ramakrishna.

In the Hindu scripture 'Devi Mahatmyam,' Mahamaya (Great Maya) is said to cover Vishnu's eyes in Yoganidra (Divine Sleep) during cycles of existence when all is resolved into one. By exhorting Mahamaya to release Her illusory hold on Vishnu, Brahma is able to bring Vishnu to aid him in killing two demons, Madhu and Kaitabh, who have manifested from Vishnu's sleeping form. Sri Ramakrishna Paramahamsa often spoke of Mother Maya and combined deep Hindu allegory with the idea that Maya is a lesser reality that must be overcome so that one is able to realize his or her true Self. Maya as Mahalaxmi was called upon when the gods and godesses were helpless against the demon Mahisasura. The combined rage of all the gods including Brahma Vishnu and Shiva created her. She is the most powerful of all the gods and godesses in heaven.

The gods gave her ornaments, weapons and her bearer, the lion. She was unassailable. She called upon the demon, had a fierce battle against Mahisasura and his huge army herself. She killed the demon and restored heaven back to the gods and godesses. Thus She is even now the protector of the universe which is lying in her lap.

Tarkshya

Trkcya is the name of a mythical being in the Rigveda, described as a horse with the epithet *aricma-nemi* "with intact wheel-rims" (RV 1.89.6, RV 10.178.1), but alternatively taken to be a bird (RV 5.51) and later identified with Garuda (Mahabharata, Harivamsha) or Garuda's father (Bhagavata Purana 6.6.2, 21), counted among the offspring of Kashyapa in Mahabharata 1.2548, 4830 and 12468.

It is also the name of the hymn RV 10.178 ascribed to *Tarkcya Aricmanemi*.

Tvastar

In Vedic religion, Tvastar (*Tvacm[*, also transliterated as Tvashtr, Tvastri or Tvashtri, nominative *Tvacm*) is the "heavenly builder", the maker of divine implements, especially Indra's vajra. He is the former of the bodies of men and animals, and therefore called "firstborn" and invoked when desiring offspring, called *garbha-pati* "lord of fetuses".

Tvastar is the Sanskrit term for "carpenter" and especially "chariot maker". Tvastar is mentioned in the Mitanni treaty, which establishes him as a Proto-Indo-Iranian divinity.

Tvastar is sometimes associated or identified with similar deities, such as Savitar, Prajapati or Pushan. He is the father of Saranyu, who twice bears twins to Vivasvat or Vayu (RV 8.26), Yama and Yami, also identified as the first humans, and the Asvins. He is also the father of Trisiras, who was killed by Indra. Tvashtri created Vrtra, a fearsome dragon, in revenge.

Tvastar is a solar deity in the Mahabharata and the Harivamsa. He is mentioned as the son of Surya and Adita in Hinduism, and said to have made the three worlds with pieces of the sun (Surya is the solar deity). Tvastar is often compared to Visvakarma.

Saranyu

Saranyu (*Saragyu*) or Saranya is the wife of Surya, and a goddess of the dawn and the clouds in Hindu mythology, and is sometimes associated with Demeter. According to Max Muller and A. Kuhn, Demeter is the mythological equivalent of the Sanskrit Saranyu, who, having turned herself into a mare, is pursued by Vivasvat, and becomes the mother of the two Asvins, the Indian Dioscuri, the Indian and Greek myths being regarded as identical. She is also the mother of Manu, Yama and Yami.

According to Farnell, the meaning of the epithet is to be looked for in the original conception of Erinys, which was that of an earth-goddess akin to Ge, thus naturally associated with Demeter, rather than that of a wrathful avenging deity.

Saragyu is the female form of the adjective *Saragyu*, meaning "quick, fleet, nimble", used for rivers and wind in the Rigveda (compare also Sarayu).

Etymologically, Saranyu may be related to Helen. In Rigveda 10.17, Saranyu is the daughter of Tvastar, and, like Helen, is abducted, and Vivasvan is given a replacement bride instead.

1ab *tvacma duhitre vahatuc kgotitidac viœvam bhuvanac sam etil*

1cd *yamasya mata paryuhyamana maho jaya vivasvato nanaua*

2ab *apaguhann amàtam martyebhya ktvi savargam adadur vivasvate*

2cd *utaœvinav abharad yat tad asid ajahad u dva mithuna Saragyu.*

Tvastar prepares the bridal of his Daughter: all the world hears the tidings and assembles.

But Yama's Mother, Spouse of great Vivasvan, vanished as she was carried to her dwelling. From mortal men they hid the Immortal Lady, made one like her and gave her to Vivasvan.

Saranyu brought to him the Asvin brothers, and then deserted both twinned pairs of children.

Hindu Mythology

Agni: Agni is the Hindu god of fire, the guardian of homes, and the protector of humans against evil.

Annapurna: In Hindu mythology, Annapurna is a goddess of foodstuff. She who causes the continual transformation of substance into energy.

Aum: Aum is the sacred mystical syllable representing the Hindu trinity.

Avatar: Avatar (Avatara) in Hindu mythology is an incarnation of a deity.

Brahma: Brahma is the Hindu supreme god and creator of the cosmos.

Chaya: In Hindu mythology, Chaya is the goddess of the beneficial effects of muted and dappled sunlight. She is the matron of metal-workers.

Durga: Durga is a Hindu deity. She is depicted as having ten arms. Her most famous exploit was slaying Mahisha. The festival of Durga puja is celebrated annually in her honour.

Dyaus: In Hindu mythology, Dyaus is the god of the sky.

Ganesha: Ganesha is an elephant headed Hindu god. He is the son of Shiva and Parvati.

Hanuman: In Hindu mythology, Hanuman is the monkey god and King of Hindustan. He assisted Rama in the recovery of his wife, Sita, from Ravana of Lanka.

Hindu: see "Hinduism"

Hinduism: Hinduism is a religion that originated in India.

Ida: In Hindu mythology, Ida is a goddess of earth, abundant food and nourishment. She is the granter of any blessings invoked through her.

Indra: Indra is the Hindu sky god. He is depicted as a four-armed man on a white elephant carrying a thunderbolt. It is he who slashes the clouds with his thunderbolt to release the rain.

Indrani: In Hindu mythology, Indrani is a voluptuously beautiful goddess. She who makes possible the perception of form through the eye.

Jagadhamba: In Hindu mythology, Jagadhamba was the goddess of the cosmic energy which motivates evolution, She Who dances the dance of life. Equivalent to the Dravidian goddess Jyestha.

Jaganath: In Hindu mythology, Jaganath was a hideously ugly god made from Krishna's bones after the god's spirit returned to heaven.

Jara: In Hindu mythology, Jara is the goddess of the household, domestic health, happiness and prosperity. The night-eater of corpses.

Kali: Kali is the Hindu goddess of destruction and death. She is the wife of Shiva.

Krishna: In Hindu mythology Krishna is the supreme god, or to some people the only god. He was the eighth avatar of Vishnu and was sent into the world to kill the demon-king Kansa who had outlawed Vishnu-worship and upset the universal balance between good and evil.

Lakshami: Lakshami is the Hindu goddess of wealth and beauty.

Lakshmi: Lakshmi is the Hindu goddess of wealth and beauty. She is the consort of Vishnu and is celebrated during Diwali.

Mahisha: Mahisha was chief of the demons in Hindu mythology. He was killed by Durga.

Manu: In Hindu mythology, Manu was the founder of the human race. He was saved by Brahma from a deluge.

Mara: In Hindu mythology, Mara is a goddess of death.

Maruts: In Hindu mythology, the Maruts are the fierce storm-beings who toss the sea into foam.

Meru: In Hindu mythology, Meru is a mountain at the North Pole, supposed to be an abode of the gods.

Parashurama: In Hindu mythology, Parashurama was the sixth avatar of Vishnu. He was the youngest son of a forest dwelling hermit with a family of sons. He was both a devout servant of the gods and a keen huntsman. In return for his prayers the gods gave him unerring skill with a bow, but demons afraid that he would use his archery against them stole his bow. Vishnu answered his prayers and entered him and gave him a battle-axe and told him to use it to fight the enemies of the gods.

While Parashurama was praying on a mountain, his mother coming home saw two people making love in a forest pool and became sexually excited. Returning home her husband detected her impure thoughts and ordered each of his sons in turn to behead her, each refused and their father punished them by making them as stupid as the cattle they looked after. When Parashurama arrived home with the battle-axe he understood his father's request, and cut off his mothers head. Immediately her head touched the floor she was restored to life cleansed of her impure thoughts, and Parashurama's father rewarded him with immunity from death in war.

Radha: In Hindu mythology, Radha is the creative, life-sustaining, auspicious, benevolent, loving and redemptive goddess of the Gopis.

Rajarajeshwari: In Hindu mythology, Rajarajeshwari is a goddess of world sustaining transcendental knowledge. She whose glances delight the universe.

Rakshasa: In Hindu mythology, the Rakshasa are male demons capabale of assuming the form of animals or humans they are completely evil, powerful creatures that delight in spreading fear, confusion, chaos, and destruction among human families and communities, finishing the trouble they cause in a murderous, ghoulish feast upon human flesh. They are among the most feared of all creatures, for they delight in mental torture of their victims.

Rakshasi: In Hindu mythology, a Rakshasi is the female form of a Rakshasa.

Ravana: In Hindu mythology, Ravana is the demon King of Lanka who abducted Sita, the wife of Rama.

Rohini: In Hindu mythology, Rohini (Red-One) is a fine-eyed goddess of stars and the rising sun. Ruler of the annual cycle and the light of knowledge.

Shakti: In Hindu mythology, Shakti is the goddess of cosmic energy, the will of evolution. It is she who gives birth to all force and all form.

Saraswati: In Hindu mythology, Saraswati (She-who-flows-eloquently-onward), is a young and beautiful four-armed goddess of fertility, procreation, purification and literature.

Savitar: In Hindu mythology, Savitar is an all-powerful sun god. He cut off one hand at a sacrifice, and priests gave him a golden hand to replace it.

Sita: In Hindu mythology, Sita is a goddess of spring, agriculture and the earth. She is the Corn Mother, the field which brings forth bounteously.

Uma: In Hindu mythology, Uma is a name of the goddess Kali.

Urmya: In Hindu and Vedic mythology, Urmya is a goddess of night and the celestial order. Protectress of sleep and she who guards against thieves.

Vac: In Hindu and Vedic mythology, Vac is the goddess of language, writing and writing-systems. She is the matron of mental endeavours and the source of voice.

Vajravarahi: In Hindu mythology, Vajravarahi (Strength-of-the-Sow) is the goddess of wisdom through experience who drives away evil spirits such as depression.

Varuna: In Hindu mythology, Varuna is a thousand-eyed god who sees all that happens in the world.

Varuni: In Hindu and Vedic mythology, Varuni was originally the waters of origin, she became the goddess of golden liquor, wine and intoxication.

Vayu: In Hindu mythology, Vayu is the god of the winds.

Visahari: In Hindu mythology, Visahari is a shape-shifting goddess of earth, uberty and the underworld. She is the controller of snakes and the spirit of mind.

Vritra: In Hindu mythology, Vritra is the snake of darkness. Enemy of Indra.

Yama: In Hindu mythology, Yama is the god of the dead, who judges and punishes souls. He is represented as of a green colour, four-armed, clad in garments of fire, crowned and seated on a buffalo. He holds a mace and noose, and drags the souls of the condemned out of the bodies to the judgement-seat.

Hindu Gods of Social Functions

The way human beings are born is the same except for the fact that the mother who gives birth suffers less or more according to her caste/class background. Our birth into a particular caste is accidental. We may have little control over our upbringing in caste-culture. After a certain age we continue to live in the culture of our own caste through a conscious decision. Having been born into a caste, very few—we can count them on our fingers—consciously move out of their caste-culture.

Though it is not in our hands to decide where we should be born, it is certainly in our own hands to decide how we should die. It is a fact that death is inevitable, but it is also a fact that such death can be moulded into a death according to our own ideas and beliefs. In the concept of death, and the experience of death, Dalitbahujans and Hindus differ in a big way. In our country there is Dalitbahujan death and there is Hindu—Brahmin, Baniya, Kshatriya and neo-Kshatriya—death. Thus, the notion of death differs between Hindus and Dalitbahujans more than it differs between people who belong to two different religions, say, Islam and Christianity.

Hindu Brahminical Death

The difference between Hindu brahminical death and Dalitbahujan death lies in the very concept of death itself. What is the Brahmin's notion of life and death? A Brahmin believes that life must be lived for the sake of death which will make him eternal. To live this way is to live a life that constantly thinks about death. Life in this universe must ensure a perennial life in the other world, that is, in heaven.

The Gods that he/she propitiates, time and again are to provide two things. One, a happy life hereon this earth, which in philosophical terms is a *kshanabhanguram* (a life that survives only a minute). At the same time, however, this short span on this earth must also be made to ensure a permanent life of privilege and pleasure. So, for a Hindu, death is a transition from this kshanabhanguram to eternity. But how does one spend this very short life here? One should eat in the name of that God who guarantees a permanent happy life. One should eat all the best things available on this earth to please the God who bestows the life of permanence. Though this body merely awaits day in and day out the transitory death that will carry it from the kshanabhanguram to eternity, it must eat rice, dal, milk, vegetables, ghee, fruit and nuts in various forms.

It must eat all this in the form of *daddoojanam, pulihoora, perugannam* (these are names of different varieties of rice foods), *pappu kuuralu,* and other curries, in several flavours. Some curries must be sour; some must be sweet; ghee plays an important role in the cuisine. The vessel holding ghee has a special place in this impermanent life. All this must be followed by sweets—laddus, jileebis, fruit salad, and so on. In the sweets too, those made of ghee occupy the highest place. In other words the kshanabhanguram body ought to be as fat as possible, with a rounded belly and unexercised muscles.

In this impermanent life, sex also plays an important role. To the Gods who enjoy the pleasures of the other world the most beautiful women like Ramba, Urvashi, Tilothama and Menaka were available. In order to facilitate this life of

Kshanabhanguram, of few pleasures, and to ensure a permanent life where more and more pleasures will be available, a son is indispensable. Since it is essential to ensure that the son is his and his alone, he needs a wife who enters his life when she is a child who has not yet attained puberty, and who will later produce a son. When a Brahmin died, if not now, in the past, to ensure his permanent pleasure in heaven, his wife must also die along with him and make herself into a sati.

In the process of working for this most-needed death, two things are necessary: (i) leisure and (ii) prayer. Leisure in this context is a divine leisure. It must keep the mind focused on acquiring the permanent life and partaking the pleasures thereof. However, this leisure is also used to develop skills that negate the Others (who are sinners because they produce the goods and commodities that prolong the life of kshanabhanguram that a Brahmin would like to end as soon as possible). Why should he eat the products produced by the labour power of Dalitbahujan castes which will prolong his life on this earth? He believes he is eating not for his own sake but eating for God who alone can ensure him moksha, or release from worldly life. Why should he indulge in sex which only results in another life, like his own, which he himself wants to end?

Even this he does for the sake of God's eternity which in turn becomes his own eternity. As I said earlier, a son is essential if he is to enter heaven, and this son is a gift of God, who is also eager to take this *punyaatma* (one who has done only good by not working at all) back to his kingdom. Why is the Brahmin a punyaatma? Because he has eaten all that is produced by the Dalitbahujans and also treated all of them as untouchable rascals. Thus, for a Brahmin, that is, for a Hindu, food and sex are two prerequisites of death.

The life of leisure when supplied with food and sex, automatically ensures two things. One, it ensures that the kshanabhanguram is longer than the unsanctified life of the Dalitbahujans because it gets the best of foods. The emphasis on sex ensures the continuance of this life in his progeny. But this truth is systematically glossed over with the second instrument—prayer. Prayer provides legitimacy for all this drama, prayer is a weapon in the hands of a Brahmin. It sets him apart from the rest of the masses. It is through this prayer that he establishes his hold over the rest of society. In a fit of madness, which might be a result of their lifetime alienation from work, life itself begins to look meaningless to them. They call this madness the life of penance.

However, when a Brahmin, who does penance throughout his life and who fathers a son to ensure a place in heaven, dies, according to the brahminical notion, it is the day when God's call comes. This death appears very different from the death of a Dalitbahujan. A Brahmin's death is adjusted with the movement of stars, *grahas* (plants), and so on. Dealth is not supposed to be mourned. Immediately after the death of a person, the Brahmins around pour into the house. Some begin pujas, some begin *prardhanas*, some begin bhajans. The people around are not allowed to weep loudly; they can only do so quietly. After the dead body is carried out, only men follow the funeral procession. Women are not allowed to take part.

From the day a Brahmin dies till the twelfth day, instead of mourning, feasting takes place. While alive the Brahmin's body is sacred (this is the reason why others must not touch him except when doing *paadapuja*, that is, touching the feet and asking for his blessing). After death that sacred soul begins its voyage to *swarga* and the body becomes untouchable (this notion got extended to Dalitbahujans also).

The priests pray for their fellow priest's soul to be given a permanent seat in the other world. Even the death of a Brahmin is a tax on the already burdened Dalitbahujans. The living Brahmins will have more feasts, which in turn will result in their extracting from the Dalitbahujan masses. In the beginning, there is a *shraadha* every day, gradually it becomes a month-wise shraaddha, and afterwards every year at the death anniversary a shraadha takes place. Even after death the soul does not merge into collectivity.

Is there any change in these notions of life and death among modern Brahmins? Not qualitatively. There are perhaps some quantitative changes, but the essence remains the same.

Even today, an average urban Brahman does not think differently from his ancestors. In post-colonial India, the Brahmins have urbanized themselves in a big way. The notion of swarga may not dominate their day-to-day life. There might be slight reversals in their notions of this life and the life after death.

Because of fast growing technology, an average Brahmin or Baniya publicly confesses that the life here should be lived with all the luxuries, but this does not mean that he or she reduces the emphasis on the permanent life in swarga. To achieve this the discourse has been changed from dharma to merit. The state and the civil society are moulded to suit 'merit' in modern times just as it was moulded to suit dharma in ancient times.

Modernized Life of Hindu (Brahmin and Baniya)

Earlier the life of kshanabhanguram involved only eating without producing the food that was eaten. Modern Brahmins and Baniyas not only eat the best and most modern food available in India, but they also have the best houses. Their styles of living have been modernized. They now have cushioned double beds, air-conditioned houses, air-cooled bedrooms (depending of course on the climate).

They have the most decorated houses. Even here modernity has been cleverly braminized. The centerpiece in a well-decorated room is not uncommonly a painting of a Hindu God or Goddess. A picture that is commonly found in the homes of brahminical families is that of Krishna's *radhasaaradhyam* (as charioteer); another popular image is that of Krishna preaching the Gita to Arjuna. All this is now couched as art and stamped with the stamp of modernity. The largest number of colour TVs are possessed by Brahmin and Baniya families.

The largest number of private cars (perhaps about 70 percent) belong to Brahmins and Baniyas. The largest number of people who travel by air in India are Brahmins and Baniyas. The highest number of people who travel in air conditioned trains and luxury buses are again Brahmins and Baniyas. Today they do all this in the name of service to the 'masses'.

They did the same thing in ancient India, calling in the service of God, done in the interest of *loka kalyanaam*. Nationalism reformulated brahminical philosophy, replacing the divine with the 'masses', as this was essential at a time when adult-franchise was on the global agenda.

The change from 'God' to 'masses' is a trick of the trade. Expanding the scope of the pleasures to be enjoyed in this world, pleasures that are increasingly on sale in a capitalist market, is an essential prerequisite, and hence the restructuring of Brahminism is the need of the day. During the post-1947 period, enjoying political power became one of the pleasures of life.

This did not diminish, however, the Hindu ambition for the permanency of life in the other world. As the Brahmins and the Baniyas exploit the Dalitbahujans from positions of political and bureaucratic powers and industrial and capital markets, there is a growing sense of sin which, in their view, may bring early death.

There is a danger that after death the sin may haunt them in the other world too. This is sought to be overcome by constructing posh temples and establishing puja rooms in their modern houses. Both in the temples and in the puja rooms of posh houses, brahminical gods—Rama, Krishna, Shiva, Ganapathi, Lakshmi, Parvathi, and so on, are depicted in modern forms. All this is to facilitate exploitation and also to prolong life here and in the other world.

Even in modern times death is mourned with feasting and feeding people from their own castes. After death, the third day celebrations, the eleventh day celebrations, month celebrations, anniversary celebrations, all are occasions for feasting. Those who have done little except cheating and eating in their lifetimes are made to be historically important persons.

Their biographies are written, their photographs are publicized and their names appear in the media. Newspaper advertisement have been modern methods of 'upper' caste celebrations of a person's death, and of the perpetuation of the dead man's memory. They have acquired crores to spend in this way. Even after a Hindu dies, living Hindus go on wasting social wealth on him. Obviously, they think this social wealth is their (Hindu) wealth.

The Dalitbahujan Death

For Dalitbahujans life now and life after death has a different meaning from that of the Hindus. For them, life is a one-time affair. The philosophy is expressed in the proverb, *puttindokasaare sachindokasaare* ('we are born only once and die only once'). A dead man/woman in a Dalitbahujan family is a loss in terms of productive work. Each person would have developed several productive skills of his/her own. Each person would have added some instruments for enhancing production.

A Kurumaa man would have discovered new areas of sheep-breeding. He would have improved the skills of cutting wool. A Kurumaa woman would have added to the skills of spinning wool. Several women spun wool, rolling the spinning stick on their thigh muscles. If a woman who discovered the techniques of wool spinning died in the process of discovery and in the process of putting it to use, sometimes the process stopped for generations. In the death of such a woman the development of technology stagnated for quite some time.

In this philosophy there is no concept of heaven or swarga. All the descriptions of swarga that are part of the day-to-day discourses of Hindu families do not exist in Dalitbahujan families. In a Dalitbahujan view, life here must be lived for life's sake. Further, life here is related to work. The more it works, the more sacred that life becomes.

The proverbs, *panee, praardhana* ('work is workship'), *panileeni paapi* ('One who does not work is a sinner') demonstrate that work alone makes life meaningful. In contrast to brahminical notions of eating, Dalitbahujans consider eating as a part of life on earth itself. Dalitbahujan women demonstrate this philosophy in which eating is considered as a part of work.

They say *anni panulu ṭiirinaayi, okka tineepani tappa* ('all my work is done; the only job left is to eat'). The routine process of eating has no relationship to God or to the sacred. As I have discussed in Chapter 1 and 2, both cooking and eating are done at a secular and mundane level.

Among the Dalitbahujans, quite a considerable debate takes place about whether human beings have to eat to live or whether they should live to eat. In their discourses they normally come to the conclusion that they have to eat to live. As work is central to their lives, every discourse relates to work. The context itself provides answers to their philosophical questions. The life of the Dalitbahujans starts with work but does not start with bed-coffee or bed-tea.

It starts with cleaning: sweeping the surroundings, washing, taking the cattle to the field, tilling the land, and son on. In their everyday lives, the question of eating comes much later. A Kurumaa's life starts straightaway, attending to the herd of sheep or entering into the field, a Goudaa starts his day by reaching the toddy tree, a Maadigaa starts his day by taking his *aare* (an instrument that cuts skin) into his hands, working at his household industry of shoe-making. Usually they do not wash their faces in the morning. Face-washing and cleaning the teeth takes place at *buvva yalla* (food eating time, around 10 a.m.). The conclusions that they have to 'eat to live' is based on their daily experience: they work first, eat later.

The act of eating is very simple. A woman quickly swallows some rice from the *buvva kunda* (rice pot) and some curry from the *kuura kanchudu* (curry bowl). She also feeds her children with the same food and carries some to her husband. The husband stops his work for a few minutes, eats his food and resumes the work based on this daily practice, they conclude that they have to eat to live.

This philosophy of eating to live is exactly the opposite of the brahminical notions of living to eat. The Gods and the Goddesses who they worship occasionally—Pochamma, Maisamma, Potaraju—do not appear to be Gods or Goddesses who order them to make obeisance in daily puja; they do not demand divine feasting. Even on festive occasions, eating does not involve even one-tenth of the ritual that the Hindu brahminical God and Goddesses require. So, the number of items of food that appear in the ritual only indicate the desire of human beings to eat those things.

At a festival, the Dalitbahujans will kill a lamb, a goat or a hen and give a small amount (paid) to the Goddess or the God once in a while, but life in terms of the prosperity or death in terms of eternity does not figure in their relationship with that Goddess or God. Fear is not totally absent, but fear is not given a philosophical justification.

Further, 'freedom from fear' does not require the eternal protection of the Goddess or the God. In other words, the notion of prayer as investment in eternal life is not the basis of the relationship between the Goddesses or Gods and the human beings. One reason for this could be that among the Dalitbahujans, of the two things that a Hindu expects as a mater of right—leisure and prayer—are absent.

The interaction of Dalitbahujans with the land, water, forests, animals and reptiles is in order to get something new out of these things. It is a scientific interaction and a creative one. This is the reason why in their day-to-day lives the earth is referred to as Mother Earth (*talli bhoodevi),* the forest is referred to as Mother Forest (*adavi talli*), the water is referred to as Mother Ganga (*Gangamma talli).*

None of these forces demand prayer; they are forces that the Dalitbahujans constantly interact with, if any of these notions exists among Hindus, it is because of the influence of Dalitization. It is through the interaction with natural forces that the new emerges, and this newness through the addition of labour (not leisure) changes into a socially useful product.

A Hindu uses every one of these socially useful products, but does not know how, or by which process, it is produced. He or she thinks that it is a product of prayer. The Dalitbahujans know that it is a product of labour. Therefore, here labour is valued and leisure is condemned.

Since this labour is combined with constant interaction with other social beings, it produces a language, a grammar and a literature, for example, a song is an integral part of labour. If any Dalitbahujan is not involved in work, such person is known as *panii paata leenoodu* (a person without work or song). The parallel proverb of brahminical Hindus is *chaduvu sandhya leenoodu* (a person without education or prayer).

In the Dalitbahujan notion of panii paata, we find creativity and science; in the brahminical notion of chaduvu sandhya, education has degenerated into a non-labouring leisure activity, which causes human beings to degenerate. Dalitbahujan philosophy is not linked to the sacred. The reason for the degeneration of Hindu society, even in the modern period, lies here. Leisure and prayer are concepts that not only have different connotations in Dalitbahujan lives, they have a very marginal role in their socio-political formation.

What does sex mean to Dalitbahujans? Sex in these lives is not an activity of leisure-based pleasure but an essential social function. This does not mean pleasure is seen as antithetical, but the brahminical leisure-based pleasure does not find any justification. A people who developed so many arts, never thought of developing, as Vatsyayana did, the arts of sexuality.

For them such sexuality has leisure-based vulgarity and it had no scope of acquiring a social base. That is the reason why the Vatsyayana type of exercise is absent even in the oral tradition of Dalitbahujans. For example, while the brahminical temples are full of sculptures based on Vatsyayana's sixty-four arts, no Dalitbahujan temple-no shrine of Pochamma, Maisamma, Kaatamaraju, Beerappa—bears such decorations.

As against a brahminical notion of sex, the Dalitbahujans perceive the man-woman pairing as a social combination of collective production and collective procreatioin, procreating children is a social function and also adds hands to the labouring process. It is not that a son does not get better treatment than a daughter, and not that a son does not play a ritual role when the father or the mother dies. *Talagooru*, a water pot carried in front of a dead body, does play a role at the time of death. But a son is essentially seen as the caretaker of old parents, not as one who ensures a place in heaven. As the notion of swarga is not central, life after death is not important. If a person lives a socially useful life, it acquires meaning.

Death is an end in itself. If someone plays a socially negative role, after death such a person becomes a devil and keeps hanging around troubling others. Thus the difference between *punyam* and *paapam* is that punyam ends life forever, while paapam turns a person into a devil. Dalitbahujan castes perform third day and eleventh day ceremonies, but after that the dead people lose their identity.

Anniversaries are not celebrated and the identity of the dead person is not retained. All the dead become part of the *peddalu* (elders who have passed way). The priest, though he comes and performs the shraaddha, never—even notionally—concedes an equal place for Dalitbahujans. No priest proclaims that in performing the shraadha for the Dalitbahujan, his or her right to enter swarga is recognized. In fact, even the death of a Dalitbahujan is a mean used by the priests to make money., In the modern urban areas, Dalitbahujans do not touch a dead body. This is because their reference point has become the Brahmin family. They have internalized Brahminism.

Concept of Death between Hindu and Dalithbahujan

Hindu and Dalitbahujan concepts of the death of women also differ. A Hindu woman does not find any important place in Hindu ritual hierarchy. A woman's death is mourned, but not eloquently.

Among 'upper' castes, when a woman dies if a man weeps loudly, such a man is said to be unmanly. But among the Dalitbahujans, weeping aloud is possible. People gather around the dead body, and generally the women outnumber the men. I remember when my mother died, my father wept. The men around began to say, 'You are a man, how can you cry like that?' But the women intervened and said, 'Let him cry, if he does not cry now when will he cry?'

It is this autonomous space and public role of women that sets the norm for man-woman relations among the Dalitbahujan castes on a different footing even at the time of death. They provide a different social space for the death of Dalitbahujan women.

After death, in the case of women too, the third day, eleventh day and the completion of one month, are observed as *Maashikam* (death ritual). After that, the dead person is only another of the dead elders. Among Dalitbahujans, the bodies of married men and women are burnt down to ashes. Children and unmarried women or single men of whatever age are buried. This practice is an extension of Brahmin practice. Cremation is an unscientific method of dealing with dead bodies because it leaves no history in the form of fossils.

If the whole world had done what the Brahmin rituals requires, the whole fossil history of human bodies would not have been available. Quite a lot of ancient and medieval history was reconstructed based on skeletons of the human that could be studies even after centuries. Burning of human bodies burns every evidence. I wish the Dalitbahujans could have developed a different practice even in this respect by giving up the practice of burning dead bodies. Today the practice of burning dead bodies is used by state agencies like the police and the army to destroy the evidence of torture and murder. Brahaminism must have evolved this practice in ancient India as the Hindus killed several Dalitbahujans who had revolted against them to destroy evidence of torture and murder.

Dalitbahujans are not used to being photographed, the old especially offer resistance. This is because the Brahmins did not allow them to maintain any history. Though architecture, painting and sculpture were Dalitbahujan occupations, a painter or an artist never left a history or a picture of his or her family. Ironically, it was the Dalitbahujans who painted and sculpted the family histories of the Brahmins and the Kshatriyas. But they were not allowed to retain a sense of their past in any visible form. This is another reason why the death of a Dalitbahujan means that he or she dies for ever. But now we must change this situation. We must see that the dead Dalitbahujans would live in the form of history; that they live in art, in paintings, in sculpture and in literature.

Hinduism left no stone unturned to destroy the wisdom, the faith, the feelings, and the images of the Dalitbahujans while they were living and also after their death. In this respect Hindu Brahminism is unparalled. Its inhumanity is unequalled. As I have shown, while living the Dalitbahujans share nothing with Hindus and even after death they share nothing with them. We should change this relationship not by Hinduizing ourselves. We must change this relatioinship by dalitizing the brahminical forces.

Throughout Indian history the Dalitbahujans have been the thesis and the brahminical forces the anti-thesis. The relationship between these forces in the form of thesis and anti-thesis has resulted in producing a synthesis but it is a mutilated synthesis. It is unnatural for a section of human beings to acquire the role of anti-thesis and continue to play the role always. It is realistic and natural that all human beings become thesis and confront nature as their anti-thesis. This is essential if the relationship between Indians as human beings is to acquire a positive homogeneity with plurality, and not a negative homogeneity which will destroy plurality itself.

Bibliography

Alaimo, Stacy : *Undomesticated Ground: Recasting Nature as Feminist Space*, Cornell University Press, Ithaca, 2000.

Appadurai, Arjun : *The Social Life of Things: Commodities in Cultural Perspective*, Cambridge University Press, New York, 1986.

Awakuni-Swetland : *Dance Lodges of the Omaha People: Building from Memory*, Routledge, London, 2001.

Barton, Craig Evan : *Sites of Memory: Perspectives on Architecture and Race,* Princeton Architectural Press, New York, 2001.

Bergson, Henri : *Matter and Memory,* G. Allen & Unwin, London, 1950.

Bigwood, Carol : *Earth Muse: Feminism, Nature, and Art,* Temple University Press, Philadelphia, 1993.

Boyer, M. Christine : *The City of Collective Memory: Its Historical Imagery and Architectural Entertainments*, MIT Press, Cambridge, 1994.

Butler, Thomas : *Memory: History, Culture and the Mind*, B. Blackwell, New York, 1989.

Campbell, John : *Past, Space, and Self,* MIT Press, Cambridge, 1994.

Carruthers, Mary : *The Book of Memory: A Study of Memory in Medieval Culture,* Cambridge University Press, New York, 1990.

Carruthers, Mary : *The Craft of Thought: Meditation, Rhetoric, and the Making of Images*, Cambridge University Press, New York, 1998.

Cicero, Marcus Tullius : *Rhetorica ad Herennium*, Harvard University Press, Cambridge, 1954.

Clanchy, M. T. : *From Memory to Written Record: England, 1066-1307,* Arnold, London, 1979.

Coleman, Janet : *Ancient and Medieval Memories: Studies in the Reconstruction of the Past*, Cambridge University Press, New York, 1992.

De Grazia, Margreta : *Subject and Object in Renaissance Culture,* Cambridge University Press, New York, 1996.

Derrida, Jacques : *Archive Fever: A Freudian Impression,* University of Chicago Press, Chicago, 1996.

Deutscher, Max : *Routledge Encyclopedia of Philosophy*, Routledge, New York, 1998.

Donald, Merlin : *Origins of the Modern Mind: Three Stages in the Evolution of Culture and Cognition*, Harvard University Press, Cambridge, 1991.

Ellul, Jacques : *The Technological Society,* Vintage Books, New York, 1964.

Engel, Susan : *Context is Everything: The Nature of Memory,* W.H. Freeman, New York, 1999.

Fara, Patricia : *Memory,* Cambridge University Press, New York, 1998.

Fentress, James : *Social Memory,* Blackwell, Cambridge, 1992.

Forty, Adrian : *The Art of Forgetting*, Berg, New York, 1999.

Foucault, Michel : *Madness and Civilization; a History of Insanity in the Age of Reason,* Pantheon Books, New York, 1965.

Gupta, Bina : *CIT: Consciousness,* Oxford University Press, New Delhi, 2003.

Harries, Karsten : *Ethical Function of Architecture*, Massachusetts. Cambridge, 1997.

Herrman, Douglas J. : *Memory in Historical Perspective: The Literature Before Ebbinghaus*, Springer-Verlag, New York, 1988.

Kearney, Richard : *Poetics of Imagining: from Husserl to Lyotard*, Harper Collins, Boston, 1991.

Klein, Norman M. : *The History of Forgetting: Los Angeles and the Erasure of Memory*, Verso, New York, 1997.

Krell, David Farrell : *Of Memory, Reminiscence, and Writing: on the Verge,* Indiana University Press, Bloomington, 1990.

LaDuke, Winona : *All Our Relations: Native Struggles for Land and Life*, South End Press, Cambridge, 1999.

Le Goff, Jacques : *History and Memory*, Columbia University Press, New York, 1992.

Leatherbarrow, David : *Uncommon Ground: Architecture, Technology, and Topography*, MIT Press, Cambridge, 2000.

Levin, David Michael : *The Body's Recollection of Being*, Routledge & Kegan Paul, Boston, 1985.

Margalit, Avishai : *The Ethics of Memory,* Harvard University Press, London, 2002.

Marshall, Joseph III : *The Lakota Way: Stories and Lessons for Living,* Viking Compass, New York, 2001.

McCulloch, W. : *Embodiments of the Mind*, MIT Press, Cambridge, 1965.

Middleton, David and Derek Edwards : *Collective Remembering*, Sage Publications, Calif., 1990.

Neisser, Ulric. : *Memory Observed: Remembering in Natural Contexts*, W.H. Freeman, San Francisco, 1982.

Otis, Laura : *Organic Memory: History And The Body In The Late Nineteenth & Early Twentieth Centuries,* University of Nebraska Press, Incoln, 1994.

Perez-Gomez, Alberto : *Architecture, Ethics and Technology*, McGill-Queens University Press, Buffalo, 1994.

Perlman, Michael : *Imaginal Memory and the Place of Hiroshima,* State University of New York Press, Albany, 1988.

Plantinga, Theodore : *How Memory Shapes Narratives: A Philosophical Essay on Redeeming the Past,* E. Mellen Press, Lewiston, 1992.

Radstone, Susannah : *Memory and Methodology*, Berg, New York, 2000.

Rossi, Paolo : *Logic and the Art of Memory: The Quest for a Universal Language,* Athlone Press, London, 2000.

Schudson, Michael : *Memory distortion: How Minds, Brains, and Societies Reconstruct the Past,* Harvard University Press, Cambridge, 1995.

Sorabji, Richard : *Aristotle on Memory,* Brown University Press, Providence, 1972.

Spector, Tom : *The Ethical Architect: The Dilemma of Contemporary Practice*, Princeton Architectural Press, New York, 2001.

Stewart, Susan : *Poetry and the Fate of the Senses,* University of Chicago Press, Chicago, 2002.

Tofts, Darren and Murray McKeich : *Memory Trade a Prehistory of Cyberculture,* Interface, North Ryde, 1998.

Trinh, T. Minh-Ha : *Woman, Native, Other: Writing Postcoloniality and Feminism,* Indiana University Press, Bloomington, 1989.

Vattimo, Gianni : *The End of Modernity: Nihilism and Hermeneutics in Post- Modern Culture,* Polity, Cambridge, 1988.

Warnock, Mary : *Memory*, Faber and Faber, London, 1987.

Weaver, Jace : *Native American Religious Identity: Unforgotten Gods,* Orbis Books, Maryknoll, 1998.

Yates, Frances A. : *The Art of Memory,* Chicago University Press, Chicago, 1966.

Young, James Edward : *The Texture of Memory: Holocaust Memorials and Meaning in Europe, Israel, and America*, Yale University Press, New Haven, 1993.

Index

S

T

U

V

W

Y

□□□